Workshop
Statistics

Springer
New York
Berlin
Heidelberg
Barcelona
Budapest
Hong Kong
London
Milan
Paris
Santa Clara
Singapore
Tokyo

Note to Instructors

For desk copies and information about supplementary materials to accompany this text, details can be found at the following addresses:

Authors' Web Sites:
http://www.dickinson.edu/~rossman/ws/index.html
http://www.piedmont.edu/math/vonoehse/workshop.html
   Materials include programs for the TI-83 graphing calculator; gopher server for data sets; Guide for Instructors, and more.
Springer-Verlag can be contacted at:
   1-800-Springer, or at http://springer-ny.com/math/text_books/tims

Workshop Statistics

Discovery with Data and the Graphing Calculator

Allan J. Rossman

DICKINSON COLLEGE

J. Barr von Oehsen

PIEDMONT COLLEGE

Springer

Library of Congress Cataloging-in-Publication Data
Rossman, Allan J.
    Workshop statistics : discovery with data and the graphing
calculator / Allan J. Rossman, J. Barr von Oehsen.
        p.   cm. — (Textbooks in mathematical sciences)
    Includes bibliographical references and index.
    ISBN 0-387-94997-6 (softcover : alk. paper)
    1. Statistics.   I. von Oehsen, J. Barr.   II. Title.
III. Series.
QA276.12.R6728   1997
519.5—dc21   97-6178

Printed on acid-free paper.

Production managed by Lesley Poliner; manufacturing supervised by Joe Quatela.
Photocomposed by University Graphics, Inc., York, PA.
Printed and bound by Hamilton Printing Co., Rensselaer, NY.
Printed in the United States of America.

9 8 7 6 5 4 3

ISBN 0-387-94997-6   Springer-Verlag New York Berlin Heidelberg   SPIN 10647040

Contents

Preface to the Calculator Version

This new version of *Workshop Statistics* responds to the interests of many instructors who do not have computers readily available for classroom use by including instructions specific to the TI-83 graphing calculator. The TI-83 is well suited for this course because of its impressive built-in statistical features, ability to interact with a computer, and programming capabilities. All of the activities and expository passages from the original text remain intact. Instructions, suggestions, and activities for the TI-83 have been interspersed throughout. Some topics and activities have been renumbered to accomodate these additions.

Throughout the text we supply enough TI-83 instructions to enable a student to complete the calculator-based activities without having to first read the *TI-83 Graphing Calculator Guidebook*. By no means are these instructions intended as a substitute for the extensive information contained in the *TI-83 Graphing Calculator Guidebook*.

Programs that have been written specifically for the TI-83, including several that allow simulations to be conducted on the calculator, and the text's data, written in TI-83 format, are available for this text. It is assumed that the necessary programs and data for each activity will be loaded into a student's calculator either by the student themselves or by the instructor at the beginning of or during the class period. Since the TI-83 does not have the memory capacity to hold all of the programs and lists, it is important to have access to either a PC or Macintosh®, and the TI-Graph Link™ software and cables.

We want to thank instructors who have provided helpful suggestions on drafts of this calculator version. These include Tom Short, Jim Bohan, Al Coons, and Joanne Schweinsberg. We also thank Charlotte Andreini of Texas Instruments for her assistance.

Acknowledgments

I want to thank Allan Rossman for sharing his yet to be published workshop statistics material with me back in 1993 when I found out that I (a pure mathematician at heart) had to teach a statistics course for the first time. I also want to thank Terry Baylor of Shippensburg University for introducing the graphing calculator to me.

My wife, Shari Prevost, deserves many thanks for all of her support and encouragement, for putting up with my many hours of calculator programming, and for introducing me to Allan Rossman. I want to thank my son Graham for all the joy he has brought into my life. Thanks also to my feline friends, Zoie, Camille, and Carl, and to my canine friend Ed.

J. BARR VON OEHSEN
February 1997

Preface

Shorn of all subtlety and led naked out of the protective fold of educational research literature, there comes a sheepish little fact: lectures don't work nearly as well as many of us would like to think.
—*George Cobb (1992)*

This book contains activities that guide students to discover statistical concepts, explore statistical principles, and apply statistical techniques. Students work toward these goals through the analysis of genuine data and through interaction with one another, with their instructor, and with technology. Providing a one-semester introduction to fundamental ideas of statistics for college and advanced high school students, *Workshop Statistics* is designed for courses that employ an interactive learning environment by replacing lectures with hands-on activities. The text contains enough expository material to stand alone, but it can also be used to supplement a more traditional textbook.

Some distinguishing features of *Workshop Statistics* are its emphases on active learning, conceptual understanding, genuine data, and the use of technology. The following sections of this preface elaborate on each of these aspects and also describe the unusual organizational structure of this text.

ACTIVE LEARNING

Statistics teaching can be more effective if teachers determine what it is they really want students to know and to do as a result of their course, and then provide activities designed to develop the performance they desire.
—*Joan Garfield (1995)*

This text is written for use with the workshop pedagogical approach, which fosters active learning by minimizing lectures and eliminating the conventional distinction between laboratory and lecture sessions. The book's activities require students to collect data, make predictions, read about studies, analyze data, discuss findings, and write explanations. The instructor's responsibilities in this setting are to check students' progress, ask and answer questions, lead class discussions, and deliver "mini-lectures" where appropriate. The essential point is that every student is actively engaged with learning the material through reading, thinking, discussing, computing, interpreting, writing, and reflecting. In this manner students construct their own knowledge of statistical ideas as they work through the activities.

The activities also lend themselves to collaborative learning. Students can work together through the book's activities, helping each other to think through the material. Some activities specifically call for collaborative effort through the pooling of class data.

The text also stresses the importance of students' communication skills. As students work through the activities, they constantly read, write, and talk with one another. Students should be encouraged to write their explanations and conclusions in full, grammatically correct sentences, as if to an educated layperson.

CONCEPTUAL UNDERSTANDING

Almost any statistics course can be improved by more emphasis on data and on concepts at the expense of less theory and fewer recipes. —David Moore (1992)

This text focuses on the "big ideas" of statistics, paying less attention to details that often divert students' attention from larger issues. Little emphasis is placed on numerical and symbolic manipulations. Rather, the activities lead students to explore the meaning of concepts such as variability, distribution, outlier, tendency, association, randomness, sampling, sampling distribution, confidence, significance, and experimental design. Students investigate these concepts by experimenting with data, often with the help of technology. Many of the activities challenge students to demonstrate their understanding of statistical issues by asking for explanations and interpretations rather than mere calculations.

To deepen students' understandings of fundamental ideas, the text presents these ideas repetitively. For example, students return to techniques of exploratory data analysis when studying properties of randomness and also in conjunction with inference procedures. They also encounter issues of data collection not just when studying randomness but also when investigating statistical inference.

GENUINE DATA

> We believe that data should be at the heart of all statistics education and that students should be introduced to statistics through data-centered courses.
> —Thomas Moore and Rosemary Roberts (1989)

The workshop approach is ideally suited to the study of statistics, the science of reasoning from data, for it forces students to be actively engaged with genuine data. Analyzing genuine data not only exposes students to the practice of statistics; it also prompts them to consider the wide applicability of statistical methods and often enhances their enjoyment of the material.

Some activities ask students to analyze data about themselves that they collect in class, while most present students with genuine data from a variety of sources. Many questions in the text ask students to make predictions about data before conducting their analyses. This practice motivates students to view data not as naked numbers but as numbers with a context, to identify personally with the data, and to take an interest in the results of their analyses.

The data sets in *Workshop Statistics* do not concentrate on one academic area but come from a variety of fields of application. These fields include law, medicine, economics, psychology, political science, and education. Many examples come not from academic disciplines but from popular culture. Specific examples therefore range from such pressing issues as testing the drug AZT and assessing evidence in sexual discrimination cases to less crucial ones of predicting basketball salaries and ranking *Star Trek* episodes.

USE OF TECHNOLOGY

> Automate calculation and graphics as much as possible.
> —David Moore (1992)

This text assumes that students have access to technology for creating visual displays, performing calculations, and conducting simulations. The preferable technology is a statistical software package, although a graphing calculator can do almost as well. Roughly half of the activities ask students to use technology. Students typically perform small-scale displays, calculations, and simulations by hand before letting the computer or calculator take over those mechanical chores.

This workshop approach employs technology in three distinct ways. First, technology performs the calculations and presents the visual displays necessary to analyze genuine data sets which are often large and cumbersome. Next, technology conducts simulations which allow students to visualize and explore the long-term behavior of sample statistics under repeated random sampling.

The most distinctive use of technology with the workshop approach is to enable students to explore statistical phenomena. Students make predictions about a particular statistical property and then use the computer to investigate their predictions, revising their predictions and iterating the process as necessary. For example, students use technology to investigate the effects of outliers on various summary statistics and the effects of sample sizes on confidence intervals.

Activities requiring the use of technology are integrated throughout the text, reinforcing the idea that technology is not to be studied for its own sake but as an indispensable tool for analyzing genuine data and a convenient device for exploring statistical phenomena.

Specific needs of the technology are to create visual displays (dotplots, histograms, boxplots, scatterplots), calculate summary statistics (mean, median, quartiles, standard deviation, correlation), conduct simulations (with binary variables), and perform inference procedures (z-tests and z-intervals for binary variables, t-tests and t-intervals for measurement variables).

ORGANIZATION

> Judge a statistics book by its exercises, and you cannot go far wrong.　　—George Cobb (1987)

For the most part this text covers traditional subject matter for a first course in statistics. The first two units concern descriptive and exploratory data analysis, the third introduces randomness and probability, and the final three delve into statistical inference. The six units of course material are divided into smaller topics, each topic following the same structure:

- *Overview:* a brief introduction to the topic, particularly emphasizing its connection to earlier topics;
- *Objectives:* a listing of specific goals for students to achieve in the topic;
- *Preliminaries:* a series of questions designed to get students thinking about issues and applications to be studied in the topic and often to collect data on themselves;
- *In-class Activities:* the activities that guide students to learn the material for the topic;

- *Homework Activities:* the activities that test students' understanding of the material and ability to apply what they have learned in the topic;
- *Wrap-up:* a brief review of the major ideas of the topic emphasizing its connection to future topics.

In keeping with the spirit of the workshop approach, hands-on activities dominate the book. Preliminary questions and in-class activities leave enough space for students to record answers in the text itself. While comments and explanations are interspersed among the activities, these passages of exposition are purposefully less thorough than in traditional textbooks. The text contains very few solved examples, further emphasizing the idea that students construct their own knowledge of statistical ideas as they work through the activities.

While the organization of content is fairly standard, unusual features include the following:

- Probability is not treated formally but is introduced through simulations. The simulations give students an intuitive sense of random variation and the idea that probability represents the proportion of times that something would happen in the long run. Because students often have trouble connecting the computer simulation with the underlying process that it models, the text first asks students to perform physical simulations involving dice and candies to help them understand the process being modeled.
- The Central Limit Theorem and the reasoning of statistical inference are introduced in the context of a population *proportion* rather than a population *mean*. A population proportion summarizes all of the relevant information about the population of a binary variable, allowing students to concentrate more easily on the concepts of sampling distribution, confidence, and significance. These ideas are introduced through physical and computer simulations which are easier to conduct with binary variables than with measurement variables. Dealing with binary variables also eliminates the need to consider issues such as the underlying shape of the population distribution and the choice of an appropriate parameter.
- Exploratory data analysis and data production issues are emphasized throughout, even in the units covering statistical inference. Most activities that call for the application of inference procedures first ask students to conduct an exploratory analysis of the data; these analyses often reveal much that the inference procedures do not. These activities also guide students to question the design of the study before drawing conclusions from the inference results. Examples used early in the text to illustrate Simpson's paradox and biased sampling reappear in the context of inference, reminding students to be cautious when drawing conclusions.

Acknowledgments

I am privileged to teach at Dickinson College, where I enjoy an ideal atmosphere for experimenting with innovative pedagogical strategies and curriculum development. I thank my many colleagues and students who have helped me in writing this book.

Nancy Baxter Hastings has directed the Workshop Mathematics Program—of which *Workshop Statistics* forms a part—with assistance from Ruth Rossow, Joanne Weissman, and Sherrill Goodlive. Barry Tesman, Jack Stodghill, Peter Martin, and Jackie Ford have taught with the book and provided valuable feedback, as have Barr von Oehsen of Piedmont College and Kevin Callahan of California State University at Hayward. Students who have contributed in many ways include Dale Usner, Kathy Reynolds, Christa Fratto, Matthew Parks, and Jennifer Becker. I also thank Dean George Allan for his leadership in establishing the productive teaching/learning environment that I enjoy at Dickinson.

I appreciate the support given to the Workshop Mathematics Program by the Fund for the Improvement of Post-Secondary Education, the U.S. Department of Education, and the National Science Foundation, as well as by Dickinson College. I also thank Springer-Verlag for their support, particularly Jerry Lyons, Liesl Gibson, and Steven Pisano. I thank Sara Buchan for help with proofreading.

Much of what I have learned about statistics education has been shaped by the writings from which I quote above. I especially thank Joan Garfield, George Cobb, Tom Short, and Joel Greenhouse for many enlightening conversations.

Finally, I thank my wonderful wife Eileen, without whose support and encouragement I would not have completed this work. Thanks also to my feline friends Eponine and Cosette.

ALLAN J. ROSSMAN
December 1995

References

Cobb, George W. (1987), "Introductory Textbooks: A Framework for Evaluation." *Journal of the American Statistical Association*, 82, 321–339.

Cobb, George W. (1992), "Teaching Statistics," in *Heeding the Call for Change: Suggestions for Curricular Action*, ed. Lynn Steen, MAA Notes Number 22, 3–43.

Garfield, Joan (1995), "How Students Learn Statistics," *International Statistical Review*, 63, 25–34.

Moore, David S. (1992), "Teaching Statistics as a Respectable Subject," in *Statistics for the Twenty-First Century*, eds. Florence and Sheldon Gordon, MAA Notes Number 26, 14–25.

Moore, Thomas L., and Rosemary A. Roberts (1989), "Statistics at Liberal Arts Colleges," *The American Statistician*, 43, 80–85.

List of Activities

NOTE: In-Class Activities appear in **boldface**.

Unit One

Exploring Data: Distributions

Topic 1:

DATA AND VARIABLES I

OVERVIEW

Statistics is the science of reasoning from *data*, so a natural place to begin your study is by examining what is meant by the term "data". The most fundamental principle in statistics is that of *variability*. Indeed, if the world were perfectly predictable and showed no variability, there would be no reason to study statistics. Thus, you will also discover the notion of a *variable* and consider different classifications of variables. You will also begin to explore the notion of the *distribution* of a set of data measuring a particular variable.

OBJECTIVES

- To begin to appreciate that *data* are numbers collected in a particular context that are studied for a purpose.
- To learn to recognize different classifications of *variables*.
- To become familiar with the fundamental concept of *variability*.
- To discover the notion of the *distribution* of a variable.
- To encounter *bar graphs* as visual displays of a distribution.
- To use the calculator to perform elementary manipulations of variables.
- To gain some exposure to the types of questions that statistical reasoning addresses.

PRELIMINARIES

1. Write a sentence describing what the word "statistics" means to you. (Here and throughout the text, please write in complete, well-constructed, grammatically correct sentences.)

2. Record in the table below the responses of students in your class to the following questions:
   • What is your gender?
   • Which of the following terms best describes your political views: liberal, moderate, or conservative?
   • Do you think that the United States should retain or abolish the penny as a coin of currency?
   • Rank your opinion of the value of statistics in society on a numerical scale of 1 (completely useless) to 9 (incredibly important).

Student	Gender	Politics	Penny	Value	Student	Gender	Politics	Penny	Value
1					13				
2					14				
3					15				
4					16				
5					17				
6					18				
7					19				
8					20				
9					21				
10					22				
11					23				
12					24				

3. How many words are in the sentence that you wrote in response to question 1?

4. Count the number of letters in each word you wrote in response to question 1. Record these below:

5. Name an area of medicine in which you suppose women physicians often choose to specialize. Also name an area in which you suppose women physicians seldom choose to specialize.

6. Take a guess as to the percentage of physicians who are women (as of 1992).

7. For each of the following pairs of sports, indicate the one you consider *more hazardous* to its participants:

   • bicycle riding and football:

   • soccer and ice hockey:

   • swimming and skateboarding:

In-Class Activities

The numbers you have recorded above are *data*. Not all numbers are data, however. Data are numbers collected in a particular context. For example, the numbers 3 and 8 do not constitute data in and of themselves. They *are* data, however, if they refer to the length (number of letters) of two different words used in answering question 1.

Since data have a context, it is very important that you refer to that context when describing a set of data. For example, it would be much more accurate to say that "the lowest ranking given to the value of statistics in society by a student in the class is 5" rather than "the smallest number in this data set is 5."

A *variable* is any characteristic of a person or thing that can be assigned a number or a category. Thus, the number of states visited by each member of your class is a variable, as is the gender of each student. A *measurement* variable is one that can assume a range of numerical values, while a *categorical* variable is one that simply records a category designation. *Binary* variables are categorical variables for which only *two* possible categories exist. These designations are quite important, for one typically employs different statistical tools depending on the type of variable measured.

The person or thing that is assigned the number or category is called the *case* or *observational unit*. In the "states visited" example above, the students in your class are the cases of interest. If we were analyzing the number of residents in each of the 50 states, the states themselves would be the cases.

Activity 1-1: Types of Variables

(a) For each of the variables listed below, indicate whether it is a measurement or a categorical variable. If it is a categorical variable, indicate whether or not it is a binary variable.

gender:

political identification:

penny question:

value of statistics:

number of states visited:

number of countries visited:

whether or not one has been to Europe:

whether or not one has been to WDW:

letters per word:

(b) Suppose that instead of recording the number of letters in each word of your sentence, you had been asked to classify each word according to the following criteria:

    1–3 letters: small word
    4–6 letters: medium word
    7–9 letters: big word
    10 or more letters: very big word

In this case, what type of variable is size of word?

(c) Suppose that instead of recording whether or not you have been to Walt Disney World, you had been asked to report the number of times that you have been to Walt Disney World. What type of variable would this have been?

As the term "variable" suggests, the values assumed by a variable differ from case to case. Certainly not every student has visited the same number of states or is of the same gender! In other words, data display *variability*. The pattern of this variability is called the *distribution* of the variable. Much of the practice of statistics concerns distributions of variables, from displaying them visually to summarizing them numerically to describing them verbally.

Activity 1-2: Penny Thoughts

(a) How many students responded to the question about whether the United States should retain or abolish the penny? How many of these voted to retain the penny? What proportion of the respondents is this?

(b) How many and what proportion of the respondents voted to abolish the penny?

(c) Create a visual display of this distribution of responses by drawing rectangles whose heights correspond to the proportions voting for each option.

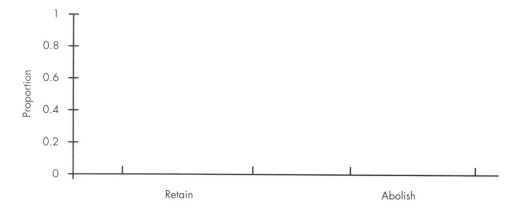

(d) Write a sentence or two describing what your analysis reveals about the attitudes of students in your class toward the penny.

The visual display you have constructed above is called a *bar graph*. Bar graphs display the distribution of categorical variable.

When a measurement variable can assume a fairly small number of possible values, one can *tally* the data by counting the *frequency* of each possible numerical response.

Activity 1-3: Entering Data into the Calculator

The first calculator operation one needs to know is entering data. On the TI-83 this is fairly straightforward. One way to enter data is to use one of the six built-in lists, L1 through L6, supplied by your calculator.

(a) Enter the data collected in question 2 of the "Preliminaries" section on the value of statistics into L1. To access L1 through L6 on your calculator follow the instructions below.

- Turn on your calculator.

- Press the STAT button.

- Press the ENTER or the 1 button .

- By using the right and left scroll buttons you can choose any of the lists.

Once you have chosen your list you can type in the data (if L1 through L6 do not show, select SetUpEditor under the STAT menu and press ENTER). For example, if the first six responses to the "value of statistics" question are 9, 9, 9, 8, 8, 9, once you enter these into L1 (you will need to press ENTER after you type each number), your screen should read as:

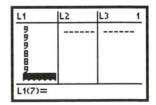

- To exit the editor, press 2nd MODE or [QUIT].

(b) Use your ⬚ 2nd ⬚ , ⬚ (⬚ , ⬚) ⬚ , and ⬚ STO▷ ⬚ buttons to enter the data collected in question 2 of the "Preliminaries" section on the value of statistics into L2. For example, if you want to enter the numbers 9, 9, 9, 8, 8, 9, into L2 from the home screen, you would type the following and then press ⬚ ENTER ⬚ .

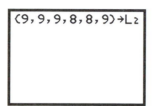

(c) Sort list L1 in ascending order using the "SortA(" feature located within your STAT menu (the "SortD(" feature sorts your list in descending order), i.e., type the following onto your home screen and then press ⬚ ENTER ⬚ :

Your L1 is now sorted in ascending order.

Activity 1-4: Value of Statistics

(a) Consider the question of students' rating the value of statistics in society on a numerical score of 1 to 9. Tally the responses by counting how many students answered 1, how many answered 2, and so on.

Rating	1	2	3	4	5	6	7	8	9
Tally (count)									

(b) By using the ALPHA button on your calculator, enter the "rating" row into a list called RATE. Type the following on your home screen and then press ENTER:

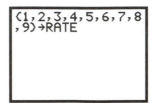

(c) Enter the "tally" row into a list named TALLY. To access this list, once you have entered it into the calculator, press 2nd and STAT. You are now in your LIST menu.

(d) Use your SetUpEditor, located in the STAT menu, to view RATE and TALLY side by side (to access these named lists go to the LIST menu). To do this, enter the following onto your home screen:

Press ENTER and then press STAT 1. You should see the lists within your STAT LIST EDITOR. You can use the SetUpEditor feature for one or more lists.

(e) Find the percent tally by dividing the TALLY list by the number of students in the class and then multiplying by 100. Use the $\boxed{\text{STO}\triangleright}$ button to enter these new data into a list named PRCNT. For example, if the class size is 35, type the following onto your home screen and press $\boxed{\text{ENTER}}$:

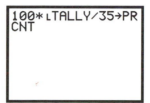

(f) Round the elements in PRCNT to the nearest 10th by using the round function located in the MATH NUM menu and reenter into PRCNT. Before pressing $\boxed{\text{ENTER}}$ your home screen should look similar to:

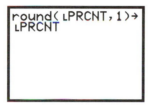

The 1 after ʟPRCNT tells the calculator to round every number in the list to 1 decimal place. If you had entered a 2 instead, then the calculator would round off to 2 decimal places, and so forth.

(g) Use the SetUpEditor feature to view the lists RATE, TALLY, and PRCNT side by side within the STAT LIST EDITOR. Identify the value(s) with the highest percentage of student responses and the value(s) with the lowest percentage of student responses.

Activity 1-5: Gender of Physicians

For each of 36 medical specialties, the table below lists the numbers of men and women physicians who identified themselves as practicing that specialty in 1992.

Specialty	Men	Women
Aerospace medicine	650	41
Allergy & immunology	2,889	552
Anesthesiology	22,978	5,170
Cardiovascular disease	15,563	915
Child psychiatry	2,981	1,637
Colon/rectal surgery	840	29
Dermatology	6,006	1,906
Diagnostic radiology	14,354	2,899
Emergency medicine	13,111	2,359
Family practice	41,261	9,708
Forensic pathology	325	98
Gastroenterology	7,438	508
General practice	18,391	2,328
General preventive medicine	877	299
General surgery	36,380	2,831
Internal medicine	67,138	18,701
Neurological surgery	4,355	146
Neurology	8,110	1,632
Nuclear medicine	1,156	216
Obstetrics/gynecology	23,497	8,090
Occupational medicine	2,465	322
Opthalmology	14,691	1,742
Orthopedic surgery	20,126	514
Otolaryngology	7,882	491
Pathology–anat./clin.	12,849	4,156
Pediatric cardiology	829	211
Pediatrics	23,842	16,573
Physical med./rehab.	3,124	1,345
Plastic surgery	4,354	334
Psychiatry	27,377	9,028
Public health	1,505	479
Pulmonary diseases	5,777	560
Radiation oncology	2,447	566
Radiology	7,064	784
Thoracic surgery	2,090	30
Urological surgery	9,290	162

(a) Lists containing the numbers of men and women physicians in each of the specialties mentioned above have been stored in a grouped file named GENPHYS.83g. To download this grouped file into your TI-83 you must use the TI-83 TI-Graph Link™ software and cable. Connect your computer to the calculator via the TI-Graph Link™ cable and then download. (You will use this method to download all files and programs from the computer). The lists, which are now located within the LIST menu of your calculator, are named MEN and WOMEN, respectively.

(b) Use the calculator to compute the percentage of women physicians in each specialty. You will need to type the following onto your home screen.

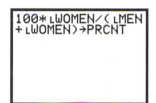

Now press ENTER. The percentage of women physicians in each specialty has been entered into a list named PRCNT.

(c) Identify the three specialties with the *highest* percentages of women physicians and the three specialties with the *lowest* percentages of women physicians.

Three highest:

Three lowest:

HOMEWORK ACTIVITIES

Activity 1-6: Types of Variables (*cont.*)

Suppose that each of the following were a variable that you were to measure for each student in your class. Indicate whether it is a measurement variable or a categorical variable; if it is categorical indicate whether it is also binary.

(a) height
(b) armspan
(c) ratio of height to armspan
(d) time spent sleeping last night
(e) whether or not the individual went to sleep before midnight last night
(f) month of birth
(g) numerical score (out of a possible 100 points) on the first exam in your course
(h) whether or not the individual scores at least 70 points on the first exam in your course
(i) distance from home
(j) whether the individual owns a credit card

Activity 1-7: Types of Variables (*cont.*)

For each of the following variables, indicate whether it is a measurement variable or a categorical (possibly binary) variable. Also identify the case (observational unit) involved. (You will encounter each of these variables later in the text.)

(a) whether a spun penny lands "heads" or "tails"
(b) the color of a Reese's Pieces candy
(c) the number of calories in a fast food sandwich
(d) the life expectancy of a nation
(e) whether an American household owns or does not own a cat
(f) the number of years that a faculty member has been at a college
(g) the comprehensive fee charged by a college
(h) for whom an American voted in the 1992 Presidential election
(i) whether or not a newborn baby tests HIV-positive
(j) the running time of an Alfred Hitchcock movie
(k) the age of an American penny
(l) the weight of an automobile
(m) whether an automobile is foreign or domestic to the United States
(n) the classification of an automobile as small, midsize, or large
(o) whether or not an applicant for graduate school is accepted
(p) the occupational background of a Civil War general
(q) whether or not an American child lives with both parents
(r) whether a college student has abstained from the use of alcohol for the past month
(s) whether or not a participant in a sport suffers an injury in a given year
(t) a sport's injury rate per 1,000 participants
(u) a state's rate of automobile thefts per 1,000 residents
(v) the airfare to a selected city from Harrisburg, Pennsylvania

(w) the average low temperature in January for a city
(x) the age of a bride on her wedding day
(y) whether the bride is older, younger, or the same age as the groom in a wedding couple
(z) the difference in ages (groom's age minus bride's age) of a wedding couple

Activity 1-8: Students' Political Views

Consider the students' self-descriptions of political inclination as liberal, moderate, or conservative.

(a) Calculate the proportion of students who identified themselves as liberal, the proportion who regard themselves as moderate, and the proportion who lean toward the conservative.
(b) Create a bar graph to display this distribution of political inclinations.
(c) Comment in a sentence or two on what your calculations and bar graph reveal about the distribution of political inclinations among these students.

Activity 1-9: Hazardousness of Sports

The following table lists the number of sports-related injuries treated in U.S. hospital emergency rooms in 1991, along with an estimate of the number of participants in the sports.

Sport	Injuries	Participants	Sport	Injuries	Participants
Basketball	646,678	26,200,000	Fishing	84,115	47,000,000
Bicycle riding	600,649	54,000,000	Horseback riding	71,490	10,100,000
Baseball, softball	459,542	36,100,000	Skateboarding	56,435	8,000,000
Football	453,684	13,300,000	Ice hockey	54,601	1,800,000
Soccer	150,449	10,000,000	Golf	38,626	24,700,000
Swimming	130,362	66,200,000	Tennis	29,936	16,700,000
Volleyball	129,839	22,600,000	Ice skating	29,047	7,900,000
Roller skating	113,150	26,500,000	Water skiing	26,633	9,000,000
Weightlifting	86,398	39,200,000	Bowling	25,417	40,400,000

(a) If one uses the number of injuries as a measure of the hazardousness of a sport, which sport is more hazardous, bicycle riding or football? soccer or ice hockey? swimming or skateboarding?

(b) Enter the data above into your calculator and compute each sport's *rate* of injuries per thousand participants.

(c) In terms of the injury rate per thousand participants, which sport is more hazardous, bicycle riding or football? soccer or ice hockey? swimming or skateboarding?

(d) How do the answers to (a) and (c) compare to each other? How do they compare to your intuitive perceptions from the Preliminaries section?

(e) List the three most and three least hazardous sports according to the injury rate per thousand participants.

(f) Identify some other factors that are related to the hazardousness of a sport. In other words, what information might you use to produce a better measure of a sport's hazardousness?

Activity 1-10: Super Bowls and Oscar Winners

Select *either* National Football League Super Bowls or movies that have won the Academy Award for Best Picture as the *cases* of interest in a study. List two measurement variables and two binary categorical variables that one might study about those cases.

Activity 1-11: Variables of Personal Interest

List three *variables* that you would be interested in studying. These can be related to anything at all and need not be things feasible for us to study in class. Be sure, however, that these correspond to thedefinition of a variable given above. Also indicate in each instance what the *case* is. Be very specific.

WRAP-UP

Since statistics is the science of *data*, this topic has tried to give you a sense of what data are and a glimpse of what data analysis entails. Data are not mere numbers: data are collected for some purpose and have meaning in some context. The guessing exercises in these activities have not been simply for your amusement; they have tried to instill in you the inclination to *think* about data in their context and to anticipate reasonable values for the data to be collected and analyzed.

You have encountered two very important concepts in this topic that will be central to the entire course: *variability* and *distribution*. You have also learned to distinguish between *measurement* and *categorical* variables. These activities have also hinted at a fundamental principle of data analysis: One should always begin analyzing data by looking at a visual display (i.e., a "picture") of the data. You have discovered a simple technique for producing such displays of categorical variables: *bar graphs*.

The next topic will continue your introduction to these basic ideas and lead you to a simple graphical display for measurement variables.

---■---

Topic 2:

DATA AND VARIABLES II

OVERVIEW

This topic continues your examination of what is meant by the terms "data" and "variable". In the previous topic you explored using graphical displays (bar graphs) for categorical variables. In this topic you will encounter a graphical display used for measurement variables. You will also continue to learn how to analyze the *distribution* of a set of data measuring a particular variable.

OBJECTIVES

- To continue exploring the notion of the *distribution* of a variable
- To encounter *dotplots* as visual displays of a distribution
- To initiate the use of the graphing calculator as an indispensable tool for analyzing real data
- To gain continued exposure to the types of questions that statistical reasoning addresses

PRELIMINARIES

1. Take a wild guess as to the number of different *states* that have been visited (or lived in) by a typical student at your school. Also guess what the fewest and most states visited by the students in your class will be.

2. Take a guess as to the proportion of students at your school who have been to Europe.

3. Place a check beside each state that you have visited (or lived in or even just driven through), and count how many states you have visited.

State	Visited?	State	Visited?	State	Visited?
Alabama		Louisiana		Ohio	
Alaska		Maine		Oklahoma	
Arizona		Maryland		Oregon	
Arkansas		Massachusetts		Pennsylvania	
California		Michigan		Rhode Island	
Colorado		Minnesota		South Carolina	
Connecticut		Mississippi		South Dakota	
Delaware		Missouri		Tennessee	
Florida		Montana		Texas	
Georgia		Nebraska		Utah	
Hawaii		Nevada		Vermont	
Idaho		New Hampshire		Virginia	
Illinois		New Jersey		Washington	
Indiana		New Mexico		West Virginia	
Iowa		New York		Wisconsin	
Kansas		North Carolina		Wyoming	
Kentucky		North Dakota			

Total states visited: _____

4. Record in the table below the following information concerning each student in your class:
- the number of states in the United States that he/she has visited;
- the number of countries that he/she has visited;
- whether or not he/she has been to Walt Disney World (WDW);
- whether or not he/she has been to Europe.

Student	Gender	States	Nations	WDW	Europe	Student	Gender	States	Nations	WDW	Europe
1						13					
2						14					
3						15					
4						16					
5						17					
6						18					
7						19					
8						20					
9						21					
10						22					
11						23					
12						24					

IN CLASS ACTIVITIES

Activity 2-1: Students' Travels

(a) Create a visual display of the distribution of the numbers of *states* visited. A horizontal scale has been drawn below; you are to place a dot for each student above the appropriate number of states visited. For repeated values, just stack the dots on top of each other.

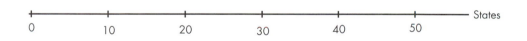

(b) Circle your own value on the display. Where do you seem to fall in relation to your peers with regard to number of states visited?

(c) Based on this display, comment on the accuracy of your guesses in the Preliminaries section.

(d) Write a paragraph of at least four sentences describing various features of the distribution of states visited. Imagine that you are trying to explain what this distribution looks like to someone who cannot see the display and has absolutely no idea about how many states people visit. Here and throughout the text, please relate your comments to the context; remember that these are states visited and not just arbitrary numbers!

The visual display that you constructed for the states visited is a *dotplot*. Dotplots are useful for displaying the distribution of small data sets of measurement variables.

Activity 2-2: Calculator Display of Students' Travels

(a) Enter into your calculator the data collected in the "Preliminaries" of the number of *states* visited by students into a list named STATE.

(b) A program named DOTPLOT.83p will enable you to create dotplots on your calculator. Download this program from the computer into your calculator.

(c) Create a dotplot of these data on your calculator. To do this press PRGM , select DOTPLOT, and then press your ENTER button twice. Input the list named STATE (which can be found in the LIST menu) at the prompt. Your screen should read as follows:

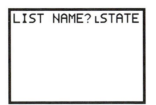

The calculator will display the dotplot after you press the ENTER button. Compare the calculator's dotplot of these data with the dotplot you created by hand above. (If there are more than 15 students in your class and no two students have the same total for the number of states visited, then your calculator's dotplot may differ from the one you created by hand). Use your TRACE button to get information about this plot.

The dotplot program turns off the coordinate axes. If you want to see the coordinate axes displayed, press 2nd ZOOM or [FORMAT] and select AxesOn.

Activity 2-3: Gender of Physicians *(cont.)*

Refer back to the data in Activity 1-5 concerning medical specialties.

(a) Download GENPHYS.83g into your calculator (if you do not have both MEN and WOMEN lists in your calculator already).

(b) Use the calculator to produce a dotplot of the distribution of percentage women. Identify a specialty that seems to have a "typical" percentage of women physicians. What percentage of this specialty's physicians are women?

(c) Based on a casual examination of this dotplot, write a brief paragraph describing key features of the distribution of percentages of women physicians.

Activity 2-4: Tennis Simulations

As part of a study investigating alternative scoring systems for the sport of tennis, researchers analyzed computer simulations of tennis matches. For 100 simulated games of tennis, the researchers recorded the number of points played in the game. (A game of tennis is won when one player wins four points with a margin of at least two points over the opponent.) The data are tallied in the following table. As examples of reading this table,

12 games ended after just four points and 2 games required eighteen points to complete.

Points in game	4	5	6	8	10	12	14	16	18
Tally (count)	12	21	34	18	9	2	1	1	2

(a) Download TENNSIM1.83l into your calculator. The data will be entered into a list named TENNI. Note that this list is not the tally, but the raw data, i.e., there are 12 fours, 21 fives, 34 sixes, and so forth. Examine a dotplot of these data with the calculator, and comment on key features of the distribution.

(b) Describe and explain the unusual granularity that the distribution exhibits.

HOMEWORK ACTIVITIES

Activity 2-5: Students' Travels *(cont.)*

Referring to the data collected above, enter the numbers of *countries* visited into the calculator and have the calculator create a dotplot of the distribution. Write a paragraph describing this distribution.

Activity 2-6: Word Lengths *(cont.)*

Refer back to lengths (number of letters) of words that you recorded back in Topic 1.

(a) Create (by hand) a dotplot of the lengths of words.
(b) Is there one particular length that occurs more often than any other? If so, what is it?
(c) Try to identify a length such that about half of your words are longer than that length and about half are shorter.
(d) Write a few sentences describing this distribution of word lengths.

Activity 2-7: Broadway Shows

The following table lists the 18 shows produced on Broadway during the week of May 31–June 6, 1993. Also listed are the type of show (play or musical), the total box office receipts generated by the show that week, the total attendance for the week, and the theater's capacity for attendance that week.

Show	Type	Receipts	Attendance	Capacity
Angels in America	Play	$326,121	6,711	7,456
Blood Brothers	Musical	$154,064	4,322	7,936
Cats	Musical	$346,723	8,386	11,856
Crazy for You	Musical	$463,377	10,016	11,720
Falsettos	Musical	$86,864	2,566	6,440
Fool Moon	Play	$163,802	5,798	10,696
The Goodbye Girl	Musical	$429,158	10,126	12,736
Guys and Dolls	Musical	$457,087	7,868	10,256
Jelly's Last Jam	Musical	$253,951	6,075	9,864
Kiss of the Spider Woman	Musical	$406,498	8,283	9,048
Les Miserables	Musical	$481,973	10,287	11,304
Miss Saigon	Musical	$625,804	13,386	14,088
Phantom of the Opera	Musical	$674,609	13,096	12,872
Shakespeare for My Father	Play	$78,898	3,291	4,520
The Sisters Rosenzweig	Play	$340,862	8,634	8,768
Someone Who'll Watch over Me	Play	$73,903	2,605	6,248
Tommy	Musical	$590,334	10,991	12,784
The Will Rogers Follies	Musical	$265,561	7,367	11,360

(a) What are the *cases* here?

(b) For each of the four variables recorded (type of show, receipts, attendance, capacity), indicate whether it is a measurement or a categorical variable. If it is a categorical variable, indicate whether or not it is a binary variable.

(c) Enter the data into the calculator and use the calculator to create a new variable representing attendance not as a raw total but rather as a percentage of the theater's capacity (divide the attendance list by the capacity list and multiply by 100). Then have the calculator produce a dotplot of the distribution of this new variable. Which show had the *highest* percentage of capacity? What was that value? Which show had the *lowest* percentage of capacity? What was that value?

(d) Write a paragraph describing some features of this distribution. (Like always, remember to relate your comments to the context.)

(e) How many productions had a higher attendance figure (in raw numbers) than *The Sisters Rosenzweig*? How many had higher receipts than *The Sisters Rosenzweig*? How many had a higher percentage of capacity than *The Sisters Rosenzweig*?

(f) Suggest in a brief sentence or two an explanation for the discrepancies among the answers to (e).

Activity 2-8: Value of Statistics *(cont.)*

Return again to the data collected in Topic 1 concerning students' ratings of the value of statistics in society.

(a) Create (either by hand or with the calculator) a dotplot of the distribution of students' responses.

(b) How many and what proportion of these students rated the value of statistics as a 5 on a 1–9 scale?

(c) How many and what proportion of these students rated the value of statistics as higher than 5?

(d) How many and what proportion of these students rated the value of statistics as less than 5?

(e) Summarize in a few sentences what the dotplot reveals about the distribution of students' ratings of the value of statistics. Specifically comment on the degree to which these students seem to be in agreement. Also address whether students seem to be generally optimistic, pessimistic, or undecided about the value of statistics.

WRAP-UP

In this topic you have continued to investigate two very important concepts: *variability* and *distribution*. You have also learned more about *measurement variables*. These activities have continued looking at a fundamental principle of data analysis, i.e., always begin analyzing data by looking at a visual display of the data. You have discovered that dotplots provide a simple technique for producing such displays of measurement variables. You have also begun to explore the utility of the graphing calculator as a tool for analyzing data.

The next topic will introduce you to some more graphical displays for displaying distributions and will also help you to develop a checklist of features to look for when describing a distribution of data.

Topic 3:

DISPLAYING AND DESCRIBING DISTRIBUTIONS

OVERVIEW

In the first topic you discovered the notion of the *distribution* of a set of data. You created visual displays (*bar graphs* and *dotplots*) and wrote verbal descriptions of some distributions. In this topic you will discover some general guidelines to follow when describing the key features of a distribution and also encounter two new types of visual displays: *stemplots* and *histograms*.

OBJECTIVES

- To develop a checklist of important features to look for when describing a distribution of data.
- To discover how to construct *stemplots* as simple but effective displays of a distribution of data.
- To learn how to interpret the information presented in *histograms*.
- To use the calculator to produce visual displays of distributions.
- To become comfortable and proficient with describing features of a distribution verbally.

PRELIMINARIES

1. Take a guess as to the length of the longest reign by a British ruler since William the Conqueror. Which ruler do you think reigned the longest?

2. What do you think is a typical age for an American man to marry? How about for a woman?

3. Take a guess as to *typical* box office receipts for a summer of 1993 movie during its first week of release.

4. Take a guess as to the box office receipts of the movie *Jurassic Park* during its *first week* of release.

5. Which would you guess is longer for most people: height or armspan (distance from fingertip to fingertip when arms are extended as far as possible)?

6. Take a guess concerning a typical foot length in centimeters for a student in your class.

7. Record below the gender, foot length, height, and armspan for the students in your class. Record the measurement variables in centimeters.

Stu	Gend	Foot length	Height	Armspan	Stu	Gend	Foot length	Height	Armspan
1					13				
2					14				
3					15				
4					16				
5					17				
6					18				
7					19				
8					20				
9					21				
10					22				
11					23				
12					24				

IN-CLASS ACTIVITIES

Activity 3-1: Hypothetical Exam Scores

Presented below are dotplots of distributions of (hypothetical) exam scores for twelve different classes. The questions following them will lead you to compile a "checklist" of important features to consider when describing distributions of data.

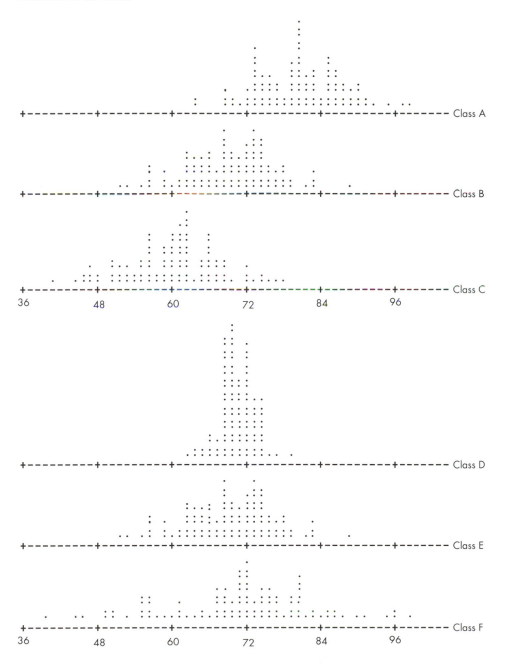

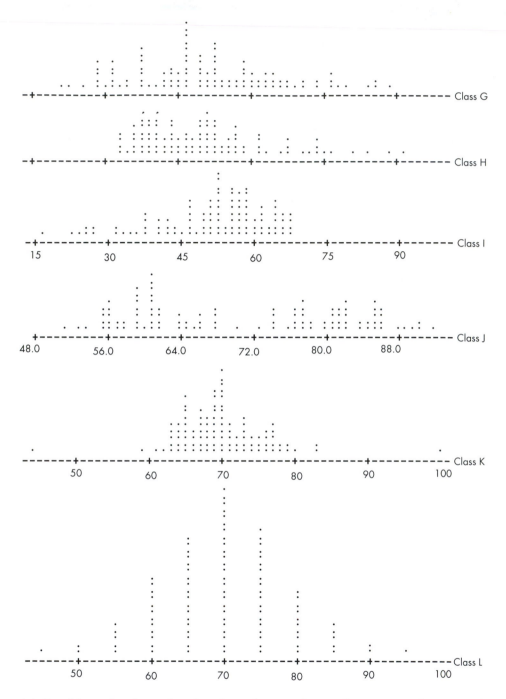

(a) Looking closely at the dotplots above, what strikes you as the most distinctive difference among the distributions of exam scores in classes A, B, and C?

(b) What strikes you as the most distinctive difference among the distributions of scores in classes D, E, and F?

(c) What strikes you as the most distinctive difference among the distributions of scores in classes G, H, and I?

(d) What strikes you as the most distinctive feature of the distribution of scores in class J?

(e) What strikes you as the most distinctive feature of the distribution of scores in class K?

(f) What strikes you as the most distinctive feature of the distribution of scores in class L?

These hypothetical exam scores illustrate six features that are often of interest when analyzing a distribution of data:

1. The *center* of a distribution is usually the most important aspect to notice and describe. Where are the data?
2. A distribution's *variability* is a second important feature. How spread out are the data?
3. The *shape* of a distribution can reveal much information. While distributions come in a limitless variety of shapes, certain shapes arise often

enough to have their own names. A distribution is *symmetric* if one half is roughly a mirror image of the other, as pictured below:

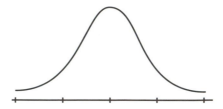

A distribution is *skewed to the right* if it tails off toward larger values:

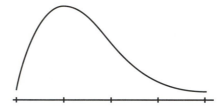

and *skewed to the left* if its tail extends to smaller values:

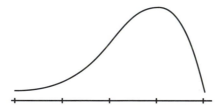

4. A distribution may have *peaks* or *clusters* which indicate that the data fall into natural subgroups.
5. *Outliers*, observations which differ markedly from the pattern established by the vast majority, often arise and warrant close examination.
6. A distribution may display *granularity* if its values occur only at fixed intervals (such as multiples of 5 or 10).

Please keep in mind that although we have formed a checklist of important features of distributions, you should not regard these as definitive rules to be applied rigidly to every data set that you consider. Rather, this list should only serve to remind you of some of the features that are *typically* of interest. Every data set is unique and has its own interesting features, some of which may not be covered by the items listed on our checklist.

Activity 3-2: British Rulers' Reigns

The table below records the lengths of reign (rounded to the nearest year) for the rulers of England and Great Britain beginning with William the Conqueror in 1066.

Ruler	Reign	Ruler	Reign	Ruler	Reign	Ruler	Reign
William I	21	Edward III	50	Edward VI	6	George I	13
William II	13	Richard II	22	Mary I	5	George II	33
Henry I	35	Henry IV	13	Elizabeth I	44	George III	59
Stephen	19	Henry V	9	James I	22	George IV	10
Henry II	35	Henry VI	39	Charles I	24	William IV	7
Richard I	10	Edward IV	22	Charles II	25	Victoria	63
John	17	Edward V	0	James II	3	Edward VII	9
Henry III	56	Richard III	2	William III	13	George V	25
Edward I	35	Henry VII	24	Mary II	6	Edward VIII	1
Edward II	20	Henry VIII	38	Anne	12	George VI	15

(a) How long was the longest reign? Who ruled the longest?

(b) What is the shortest reign? Who ruled the shortest time? What do you think this value really means?

One can create a visual display of the distribution called a *stemplot* by separating each observation into two pieces: a "stem" and a "leaf." When the data consist primarily of two-digit numbers, the natural choice is to make the tens digit the stem and the ones digit the leaf. For example, a reign of 21 years would have 2 as the stem and 1 as the leaf; a reign of 2 years would have 0 as the stem and 2 as the leaf.

(c) Fill in the stemplot below by putting each leaf on the row with the corresponding stem. We have gotten you started by filling in the reign lengths of William I (21 years), William II (13 years), Henry I (35 years), and Stephen (19 years).

```
0 |
1 | 3 9
2 | 1
3 | 5
4 |
5 |
6 |
```

(d) The final step to complete the stemplot is to order the leaves (from smallest to largest) on each row. Reproduce the stemplot below with the leaves ordered.

(e) Write a short paragraph describing the distribution of lengths of reign of British rulers. (Keep in mind the checklist of features that we derived above. Also please remember to relate your comments to the context.)

The *stemplot* is a simple but useful visual display. Its principal virtues are that it is easy to construct by hand (for relatively small sets of data) and that it retains the actual values of the observations. Of particular convenience is the fact that the stemplot also *sorts* the values from smallest to largest.

(f) Find a value such that one-half of these 40 British rulers enjoyed reigns longer than that value and one-half of them ruled for fewer years than that value. (Make use of the fact that the stemplot has arranged the values in order.)

(g) Find a value such that one-quarter of these 40 rulers enjoyed longer reigns and three-quarters of them had shorter reigns than that value.

(h) Find a value such that three-quarters of these 40 British rulers ruled for more years, and one-quarter of them ruled for fewer years than that value.

A rough but handy summary of a distribution is obtained by these three numbers together with the minimum and maximum. These values comprise the *five-number summary* of a distribution. You will learn to calculate this summary more formally in upcoming topics.

The *histogram* is a visual display similar to a stemplot but which can be produced even with very large data sets; it also permits more flexibility than does the stemplot. One constructs a histogram by dividing the range of the data into subintervals of equal length, counting the number (also called the *frequency*) of observations in each subinterval, and constructing rectangles whose heights correspond to the counts in each subinterval. Equivalently, one could make the rectangle heights correspond to the proportions (also called the *relative frequency*) of observations in the subintervals.

Activity 3-3: Pennsylvania College Tuitions

The October 23, 1991, issue of *The Chronicle of Higher Education* listed 1991–92 tuition and fee charges for a large number of American colleges and universities. The following histogram represents this information for the 157 Pennsylvania colleges listed. (Notice that the calculator software reports the midpoint of each subinterval. Thus, for example, the first rectangle indicates that 20 colleges had tuitions between $1000 and $3000.)

Tuition   N = 157

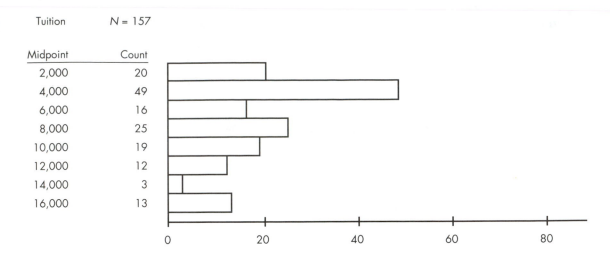

Midpoint	Count
2,000	20
4,000	49
6,000	16
8,000	25
10,000	19
12,000	12
14,000	3
16,000	13

(a) How many of the colleges had tuitions between $5,000 and $7,000?

(b) How many colleges had tuitions greater than $13,000? What proportion of the colleges listed had tuitions greater than $13,000?

(c) Is the distribution roughly symmetric or skewed in one direction or the other?

(d) How many distinct clusters (or peaks) can you identify in the distribution? Roughly where do they fall? Can you come up with a reasonable explanation of which kinds of colleges tend to fall in which clusters?

Activity 3-4: Creating Histograms on the Calculator

The next basic operation on the TI-83 that you need to understand is how to create visual displays using STAT PLOT. Before attempting to create a histogram (or any other statistical display) you must make sure that your calculator has all graphing functions turned off. To access the STAT PLOT menu, press   2nd   Y= . The TI-83 screen should look similar to:

The STAT PLOTS menu above may look different from yours since there are many possible combinations of statistical plots. The window above tells you that Plot 1 is off and it is set up to do a histogram of L1, Plot 2 is off and it is set up to do a scatterplot of L2 versus L1, and Plot 3 is off and it is set up to do a boxplot of L5 (you will learn about boxplots and scatterplots later).

(a) Enter the data of the British rulers' reigns of Activity 3-2 into a list named REIGN.

(b) Access the STAT PLOT menu and press $\boxed{\text{ENTER}}$ or $\boxed{1}$.

Set up your Plot1 as:

This screen tells you that Plot1 is on, that you have six choices of graphical displays (scatterplot, xyline, histogram, modified boxplot, boxplot, and normal probability plot), the capability to choose any list, and a choice for frequency. Since the histogram is highlighted, Plot1 is ready to display a histogram of REIGN. (The "1" placed after the frequency prompt tells the calculator to count each exam score only once).

(c) Adjust the viewing window for the histogram. For example, after pressing the $\boxed{\text{WINDOW}}$ button, you might input the following numbers.

```
WINDOW
 Xmin=0
 Xmax=74
 Xscl=10.5
 Ymin=0
 Ymax=14
 Yscl=1
 Xres=1
```

(d) Press the GRAPH button and write a few sentences describing this distribution.

A nice feature of the TI-83 is that it can give you the frequency and the boundary of each class in your histogram by pressing the TRACE button and then the right and left scroll buttons. For example, the graphic below tells you that eight rulers reigned between min = 21 years and max < 31.5 years.

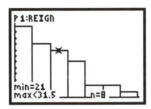

(e) Go back to your window settings and experiment with different widths of the histogram's subintervals. For example, try 5 or 15 as alternative values for the calculator's Xscl. You may find that one width will give you a better understanding of your data as opposed to another. How has the histogram changed with your new values?

Activity 3-5: Students' Measurements

Consider the data on students' physical measurements that you gathered above. Enter the foot lengths into the calculator, and have the calculator produce a dotplot and histogram of the foot lengths. Write a paragraph commenting on key features of this distribution. (Please remember always to relate such comments to the context. Imagine that you are describing this distribution of foot lengths to someone who has absolutely no idea how long college students' feet are. By all means think about the checklist of features that we have developed as you write this paragraph.)

HOMEWORK ACTIVITIES

Activity 3-6: Hypothetical Manufacturing Processes

Suppose that a manufacturing process strives to make steel rods with a diameter of 12 centimeters, but the actual diameters vary slightly from rod to rod. Suppose further that rods with diameters within ± 0.2 centimeters of the target value are considered to be within specifications (i.e., acceptable). Suppose that 50 rods are collected for inspection from each of four processes and that the dotplots of their diameters are as follows:

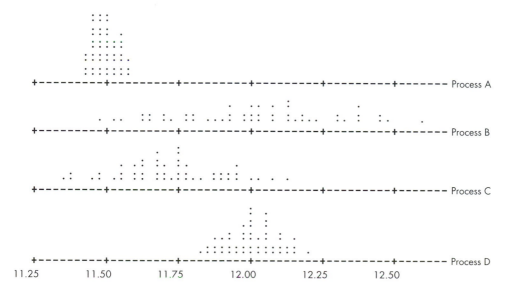

Write a paragraph describing each of these distributions, concentrating on the center and variability of the distributions. Also address each of the following questions in your paragraph:

- Which process is the best as is?
- Which process is the most stable, i.e., has the least variability in rod diameters?
- Which process is the least stable?
- Which process produces rods whose diameters are generally farthest from the target value?

Activity 3-7: Marriage Ages

Listed below are the ages of a sample of 24 couples taken from marriage licenses filed in Cumberland County, Pennsylvania in June and July of 1993.

Couple	Husband	Wife	Couple	Husband	Wife	Couple	Husband	Wife
1	25	22	9	31	30	17	26	27
2	25	32	10	54	44	18	31	36
3	51	50	11	23	23	19	26	24
4	25	25	12	34	39	20	62	60
5	38	33	13	25	24	21	29	26
6	30	27	14	23	22	22	31	23
7	60	45	15	19	16	23	29	28
8	54	47	16	71	73	24	35	36

(a) Select *either* the husbands' ages or the wives' ages, and construct (by hand) a stemplot of their distribution. (Indicate which spouse you are analyzing.)

(b) Write a short paragraph describing the distribution of marriage ages for whichever spouse you chose.

Activity 3-8: Hitchcock Films

The following table lists the running times (in minutes) of the videotape versions of 22 movies directed by Alfred Hitchcock.

Film	Time	Film	Time
The Birds	119	Psycho	108
Dial M for Murder	105	Rear Window	113
Family Plot	120	Rebecca	132
Foreign Correspondent	120	Rope	81
Frenzy	116	Shadow of a Doubt	108
I Confess	108	Spellbound	111
The Man Who Knew Too Much	120	Strangers on a Train	101
Marnie	130	To Catch a Thief	103
North by Northwest	136	Topaz	126
Notorious	103	Under Capricorn	117
The Paradine Case	116	Vertigo	128

(a) Construct a stemplot of this distribution.

(b) Comment on key features of this distribution.

(c) One of these movies is particularly unusual in that all of the action takes place in one room and Hitchcock filmed it without editing. Explain how you might be able to identify this unusual film based on the distribution of the films' running times.

Activity 3-9: *Jurassic Park* Dinosaur Heights

In the blockbuster movie *Jurassic Park*, dinosaur clones run amok on a tropical island intended to become mankind's greatest theme park. In Michael Crichton's novel on which the movie was based, the examination of dotplots of dinosaur heights provides the first clue that the dinosaurs are not as controlled as the park's creator would like to believe. Here are reproductions of two dotplots presented in the novel:

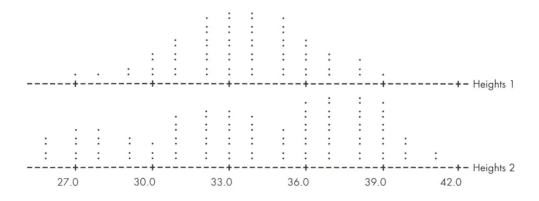

(a) Comment briefly on the most glaring difference in these two distributions of dinosaur heights.

(b) The cynical mathematician Ian Malcolm (a character in the novel) argues that one of these distributions is characteristic of a normal biological population, while the other is what one would have expected from a controlled population which had been introduced in three separate batches (as these dinosaurs had). Identify which distribution corresponds to which type of population.

(c) Take a closer look at the first distribution. There is something suspicious about it that suggests that it does not come from real data but rather from the mind of an author. Can you identify its suspicious quality?

Activity 3-10: Summer Blockbuster Movies

For each of the 43 movies released by major motion picture studios between the Memorial Day and Labor Day weekends of 1993, the box office incomes during its first and second weeks of release (in millions of dollars, taken from various issues of *Variety*) are displayed in the table below.

Movie	$ Week 1	$ Week 2
Cliffhanger	25.021	16.983
Made in America	14.104	10.193
Super Mario Brothers	9.675	5.489
Menace II Society	4.953	4.530
Guilty as Sin	7.783	5.190
Life with Mikey	4.820	3.178
Jurassic Park	79.218	61.691
What's Love Got to Do with It?	1.897	5.767
Last Action Hero	20.833	11.718
Sleepless in Seattle	27.147	21.727
Dennis the Menace	16.084	13.890
The Firm	49.239	28.979
Snow White (reissue)	12.615	9.881
Son-in-Law	9.984	7.697
In the Line of Fire	23.003	19.903
Rookie of the Year	15.289	11.849
Weekend at Bernie's II	5.936	3.516
Hocus Pocus	13.200	8.754
Free Willy	14.001	14.307
Poetic Justice	15.903	7.102
Coneheads	10.928	5.707
Another Stakeout	8.163	5.460
Robin Hood: Men in Tights	3.170	12.112

Movie	$ Week 1	$ Week 2
Rising Sun	22.086	12.791
So I Married an Axe Murderer	5.588	2.764
Tom and Jerry: The Movie	1.985	0.941
The Fugitive	37.618	33.360
The Meteor Man	3.918	2.060
My Boyfriend's Back	2.319	0.800
Searching for Bobby Fischer	0.401	1.847
Hard Target	13.497	6.887
Surf Ninjas	2.991	1.038
Manhattan Murder Mystery	2.604	2.232
The Man Without a Face	1.889	6.112
Needful Things	7.113	4.658
Only the Strong	1.775	0.930
Father Hood	1.774	1.073
Son of the Pink Panther	1.604	0.714
The Thing Called Love	0.556	0.364
Fortress	4.728	1.589
Calendar Girl	2.011	0.426
Kalifornia	1.435	0.546
Boxing Helena	0.958	0.378

(a) How much did *Jurassic Park* make in its first week? How does this compare with your guess from the "Preliminaries" section?

(b) Select two other movies from the list (preferably ones that you have seen), and record their names and first week incomes.

(c) Use the calculator to produce a dotplot of the distribution of first week's box office income. What (roughly) is a typical income? Where (roughly) do *Jurassic Park* and your other two movies fall in the distribution?

(d) Use the calculator to produce a new variable: *change* in box office income from the first week to the second. Look at a dotplot of this distribution. What (roughly) is a typical change? Where (roughly) do *Jurassic Park* and your other two movies fall?

(e) Use the calculator to produce yet another new variable: *percentage change* in box office income from the first week to the second. Look at a dotplot of this distribution, and write a few sentences commenting on its key features.

(f) What (roughly) is a typical percentage change? Where (roughly) do *Jurassic Park* and your other two movies fall?

(g) Explain why *Jurassic Park* has one of the two worst drop-offs in income, yet has a roughly typical percentage change in income from the first to the second week.

Activity 3-11: Turnpike Distances

The Pennsylvania Turnpike extends from Ohio in the west to New Jersey in the east. The distances (in miles) between its exits as one travels west to east are listed in the table.

Exit	Name	Miles	Exit	Name	Miles
1	Ohio Gateway	*	16	Carlisle	25
1A	New Castle	8	17	Gettysburg Pike	9.8
2	Beaver Valley	3.4	18	Harrisburg West Shore	5.9
3	Cranberry	15.6	19	Harrisburg East	5.4
4	Butler Valley	10.7	20	Lebanon-Lancaster	19
5	Allegheny Valley	8.6	21	Reading	19.1
6	Pittsburgh	8.9	22	Morgantown	12.8
7	Irwin	10.8	23	Downingtown	13.7
8	New Stanton	8.1	24	Valley Forge	14.3
9	Donegal	15.2	25	Norristown	6.8
10	Somerset	19.2	26	Fort Washington	5.4
11	Bedford	35.6	27	Willow Grove	4.4
12	Breezewood	15.9	28	Philadelphia	8.4
13	Fort Littleton	18.1	29	Delaware Valley	6.4
14	Willow Hill	9.1	30	Delaware River Bridge	1.3
15	Blue Mountain	12.7			

(a) In preparation for constructing a histogram to display this distribution of distances, count how many values fall into each of the subintervals listed in the table below:

Miles	0.1–5.0	5.1–10.0	10.1–15.0	15.1–20.0
Tally (count)				
Miles	20.1–25.0	25.1–30.0	30.1–35.0	35.1–40.0
Tally (count)				

(b) Construct (by hand) a histogram of this distribution.

(c) Comment in a few sentences on key features of this distribution.

(d) Find a value such that half of the exits are more than this value apart and half are less than this value. Also explain why such a value is not unique.

(e) If a person has to drive between consecutive exits and only has enough gasoline to drive 20 miles, is she very likely to make it? Assume that you do not know *which* exits she is driving between. Explain your answer.

(f) Repeat (e) if she only has enough gasoline to drive 10 miles.

Activity 3-12: College Tuition Increases

The following table lists the percentage increase in Dickinson College's comprehensive fee (tuition, room, and board) between 1971 and 1994. For example, the first entry indicates that the comprehensive fee rose 10.0% between the 1970–71 and 1971–72 academic years.

Year	1971	1972	1973	1974	1975	1976	1977	1978
% increase	10.0	5.2	6.7	3.8	7.9	7.4	6.2	6.7
Year	1979	1980	1981	1982	1983	1984	1985	1986
% increase	6.7	9.1	16.6	12.6	13.6	8.1	9.9	10.1
Year	1987	1988	1989	1990	1991	1992	1993	1994
% increase	9.7	7.8	9.5	7.7	7.2	6.9	5.8	4.9

(a) Create a stemplot of the distribution of percentage increases.

(b) Calculate the five-number summary of these percentage increases.

(c) Comment briefly on the validity of each of these statements:

- Most of these 24 years had a percentage increase of at least 10%.
- Both 1993 and 1994 had percentage increases in the lower quarter of this distribution.
- The percentage increase in 1981 was clearly an outlier compared to the other percentage increases over these years.
- More than three-fourths of these years had percentage increases of greater than 6%.
- Half of these 24 years had percentage increases between 6% and 10%.

(d) Does the stemplot by itself allow you to say anything about trends in these percentage increases over time? Explain.

Activity 3-13: ATM Withdrawals

The following table lists both the number and total amount of cash withdrawals from an automatic teller machine (ATM) during each month of 1994. (For example, January saw nine withdrawals which totaled $1020.)

Month	#	Total	Month	#	Total	Month	#	Total
January	9	$1020	May	8	$980	September	10	$850
February	8	$890	June	13	$1240	October	10	$1010
March	10	$970	July	4	$750	November	7	$860
April	9	$800	August	9	$1130	December	14	$1680

(a) Create a dotplot of the distribution of the *number* of withdrawals in each month, and comment briefly on the distribution.
(b) Create a dotplot of the distribution of the *total* amount withdrawn in each month, and comment briefly on the distribution.
(c) Which month had the most withdrawals and by far the most cash withdrawn? Suggest an explanation for this.
(d) This individual took two extended trips in one of these months. Can you guess which month it was based on these data? Explain.

Activity 3-14: Voting for Perot

The following table lists by state (and D.C.) the percentage of the popular vote received by Ross Perot in the 1992 Presidential election.

State	Region	% Perot	State	Region	% Perot
Alabama	South	10.8	Montana	West	26.4
Alaska	West	27.6	Nebraska	Midwest	23.7
Arizona	West	24.1	Nevada	West	26.7
Arkansas	South	10.6	New Hampshire	Northeast	22.8
California	West	20.8	New Jersey	Northeast	15.9
Colorado	West	23.5	New Mexico	West	16.4
Connecticut	Northeast	21.7	New York	Northeast	15.8
Delaware	South	20.6	North Carolina	South	13.7
Dist. of Colum.	South	4.2	North Dakota	Midwest	23.2
Florida	South	19.9	Ohio	Midwest	21.0
Georgia	South	13.3	Oklahoma	South	23.1
Hawaii	West	14.4	Oregon	West	25.1
Idaho	West	27.8	Pennsylvania	Northeast	18.3
Illinois	Midwest	16.9	Rhode Island	Northeast	22.8
Indiana	Midwest	19.9	South Carolina	South	11.6
Iowa	Midwest	18.8	South Dakota	Midwest	21.8
Kansas	Midwest	27.4	Tennessee	South	10.1
Kentucky	South	13.8	Texas	South	22.2
Louisiana	South	12.0	Utah	West	28.8
Maine	Northeast	30.4	Vermont	Northeast	22.6
Maryland	South	14.4	Virginia	South	13.7
Massachusetts	Northeast	22.9	Washington	West	24.3
Michigan	Midwest	19.2	West Virginia	South	15.9
Minnesota	Midwest	24.1	Wisconsin	Midwest	21.7
Mississippi	South	8.8	Wyoming	West	25.6
Missouri	Midwest	21.8			

Use the calculator to create visual displays (dotplot, histogram) of these percentages of Perot votes. (Ignore for now the different regions of the country.) Write a paragraph describing your findings about the distribution of Perot vote percentages.

Activity 3-15: Word Lengths *(cont.)*

Reconsider the data that you collected in Topic 1 concerning the number of letters in your words. Comment on features of this distribution with regard to the six features enumerated above.

WRAP-UP

With this topic you have progressed in your study of distributions of data in many ways. You have created a checklist of various features to consider when describing a distribution verbally: *center, spread, shape, clusters/peaks, outliers, granularity*. You have encountered three shapes that distributions often follow: *symmetry, skewness to the left*, and *skewness to the right*. You have also discovered two new techniques for displaying a distribution: *stemplots* and *histograms*.

Even though we have made progress by forming our checklist of features, we have still been somewhat vague to this point about describing distributions. The next two topics will remedy this ambiguity somewhat by introducing you to specific numerical measures of certain features (namely, *center* and *spread*) of a distribution. They will also lead you to studying yet another visual display: the *boxplot*.

Topic 4:

MEASURES OF CENTER

OVERVIEW

You have been exploring distributions of data, representing them graphically, and describing their key features verbally. For convenience, it is often desirable to have a single numerical measure to summarize a certain aspect of a distribution. In this topic you will encounter some of the more common measures of the center of a distribution, investigate their properties, apply them to some genuine data, and expose some of their limitations.

OBJECTIVES

- To learn to calculate certain statistics (*mean, median, mode*) for summarizing the center of a distribution of data.
- To investigate and discover properties of these summary statistics.
- To explore the statistical property of *resistance* as it applies to these statistics.
- To develop an awareness of situations in which certain measures are and are not appropriate.
- To recognize that these numerical measures do not summarize a distribution completely.
- To acquire the ability to expose faulty conclusions based on misunderstanding of these measures.

PRELIMINARIES

1. Take a guess as to how long a typical member of the U.S. Supreme Court has served.

2. Take a guess concerning the distance from the Sun for a typical planet in our solar system.

3. Make guesses about the closest and farthest a student in your class is from "home." Also guess the distance from home of a typical student in the class.

4. Record in the table below the distances from home (estimated in miles) for each student in the class.

Student	Distance	Student	Distance	Student	Distance
1		9		17	
2		10		18	
3		11		19	
4		12		20	
5		13		21	
6		14		22	
7		15		23	
8		16		24	

IN-CLASS ACTIVITIES

Activity 4-1: Supreme Court Service

The table following lists the justices comprising the Supreme Court of the United States as of October 1994. Also listed is the year of appointment and the tenure (years of service) for each.

Supreme Court Justice	Year	Tenure
William Rehnquist	1972	22
John Paul Stevens	1975	19
Sandra Day O'Connor	1981	13
Antonin Scalia	1986	8
Anthony Kennedy	1988	6
David Souter	1990	4
Clarence Thomas	1991	3
Ruth Bader Ginsburg	1993	1
Stephen Breyer	1994	0

(a) Create a dotplot of the distribution of these years of service.

(b) What number might you choose if you were asked to select a single number to represent the *center* of this distribution? Briefly explain how you arrive at this choice.

We will consider three commonly used measures of the *center* of a distribution:

- The *mean* is the ordinary arithmetic average, found by summing (adding up) the values of the observations and dividing by the number of observations.
- The *median* is the middle observation (once they are arranged in order); you will construct a formula that will help you to calculate medians.
- The *mode* is the most common value; i.e., the one that occurs most frequently.

(c) Calculate the *mean* of these years of service. Mark this value on the dotplot above with an "x."

(d) How many of the nine people have served more than the mean number of years? How many have served less than the mean number of years?

(e) Calculate the *median* of these years of service. Mark this value on the dotplot above with an "o."

(f) How many of the nine people have served more than the median number of years? How many have served less than the median number of years?

It is easy enough to pick out the median (the middle observation) in a small set of data, but we will try to come up with a general rule for finding the *location* of the median. The first step, of course, is to arrange the observations *in order* from smallest to largest. Let *n* denote the *sample size*, the number of observations in the data set.

(g) With the data you analyzed above (where $n = 9$), the median turned out to be which ordered observation (the second, the third, the fourth, . . .)?

(h) Suppose that there had been $n = 5$ observations; the median would have been which (ordered) one? What if there had been $n = 7$ observations? How about if $n = 11$? How about if $n = 9$? What about $n = 13$?

$n = 5$: $n = 11$:
$n = 7$: $n = 13$:
$n = 9$:

(i) Try to discover the pattern in the question above to determine a general formula (in terms of the sample size *n*) for finding the *location* of the median of an odd number of ordered observations.

Activity 4-2: Faculty Years of Service

The following table presents the years of service of eight college professors:

Professor	Yrs	Professor	Yrs	Professor	Yrs	Professor	Yrs
Baric	31	Hastings	7	Reed	1	Stodghill	28
Baxter	15	Prevost	3	Rossman	6	Tesman	6

(a) Rewrite these values in order from smallest to largest.

(b) Calculate the mean of these years of service.

(c) Determine the mode of these years of service.

(d) Explain why finding the median is less straightforward in this case than in the case of nine Supreme Court justices.

If there are an even number of observations, the median is defined to be the average (mean) of the middle two observations.

(e) With this definition in mind, calculate the median of these years of service.

These activities should have led you to discover that the median of an odd number of observations can be found in position $(n + 1)/2$ (once the values are arranged in order), while the median of an even number of observations is the mean of those occupying positions $n/2$ and $(n/2) + 1$.

More important than the ability to calculate these values is understanding their properties and interpreting them correctly.

(f) Enter the faculty years of service into your calculator.

(g) Use your calculator to compute the mean and median of these years of service by selecting 1-Var Stats from the STAT CALC menu. For example, if you named your list above as FACUL, then you would type the following onto your home screen and then press ENTER :

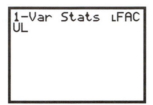

Notice the arrow in the lower left hand corner of your screen. This arrow lets you know that you can scroll down to view more information. Use the scroll buttons on your calculator to view all of the information. Write the mean $\bar{x}$ and median below.

Activity 4-3: Properties of Averages

Reconsider the hypothetical exam scores whose distribution you analyzed in Activity 3-1. (The data appear in Appendix A.)

(a) Based on the distributions as revealed in the dotplots in Activity 2-1, how would you expect the mean exam score to compare among classes A, B, and C? What about the median exam scores in these three classes?

(b) Lists containing the hypothetical exam scores have been stored as a grouped file named HYPOFEAT.83g. Download this grouped file into your calculator. The lists have been named CLSA, CLSB, CLSC, CLSD,

CLSE, CLSF, CLSG, CLSH, and CLSI, respectively. Use the calculator to compute the mean and median exam scores in classes A, B, and C. Record the results below. Do the calculations confirm your intuitions expressed in (a)?

	Class A	Class B	Class C
Mean			
Median			

(c) Consider class G. Do you expect the mean and median of this distribution to be close together, do you expect the mean to be noticeably higher than the median, or do you expect the median to be noticeably higher than the mean?

(d) Repeat (c) in reference to class H.

(e) Repeat (c) in reference to class I.

(f) Use the calculator to compute the mean and median exam scores in classes G, H, and I. Record the results below, and indicate whether the calculations confirm your intuitions expressed in (c), (d), and (e).

	Class G	Class H	Class I
Mean			
Median			

(g) Summarize what classes G, H, and I indicate about how the shape of the distribution (symmetric or skewed in one direction or the other) relates to the relative location of the mean and median.

To investigate the effect of outliers on these measures of center, reconsider the data from Activity 4-1 on Supreme Court justices' years of tenure.

(h) Enter the data into the calculator and compute the mean and median of the years of tenure. Record the results in the first row of the table below.

(i) Now imagine that Justice Rehnquist has served for 42 years rather than 22. Have the calculator recompute the mean and median in this case; record the values in the table.

(j) Finally, suppose that Justice Rehnquist's 22 years of service had been mistakenly recorded as 222 years. Again have the calculator recompute the mean and median, recording the values in the table.

	Mean	Median
Justices		
Justices with "big" outlier		
Justices with "huge" outlier		

A measure whose value is relatively unaffected by the presence of outliers in a distribution is said to be *resistant*.

(k) Based on these calculations, would you say that the mean or the median is resistant? Based on the definition of each, explain briefly why it is or is not resistant.

(l) Is there any limit in principle to how large or small the mean can become just by changing *one* of the values in the distribution?

(m) Does it make sense to talk about the mean gender of the students in your class? How about the median gender? Is the mode of the genders a sensible notion?

One can calculate the mean only with measurement variables. The median can be found with measurement variables and with categorical variables for which a clear ordering exists among the categories. The mode applies to all categorical variables but is only useful with some measurement variables.

Activity 4-4: Readability of Cancer Pamphlets

Researchers in Philadelphia investigated whether pamphlets containing information for cancer patients are written at a level that the cancer patients can comprehend. They applied tests to measure the reading levels of 63 cancer patients and also the readability levels (based on such factors as lengths of sentences and numbers of polysyllabic words) of 30 cancer pamphlets (based on such factors as the lengths of sentences and number of polysyllabic words). These numbers correspond to grade levels, but patient reading levels of under grade 3 and above grade 12 are not determined exactly.

The tallies in the following table indicate the number of patients at each reading level and the number of pamphlets at each readability level. The dotplots below display these distributions with "under 3" displayed as 2 and "above 12" displayed as 13 for convenience.

Patients' reading level	Tally
Under 3	6
3	4
4	4
5	3
6	3
7	2
8	6
9	5
10	4
11	7
12	2
Above 12	17

Pamphlets' readability	Tally
6	3
7	3
8	8
9	4
10	1
11	1
12	4
13	2
14	1
15	2
16	1

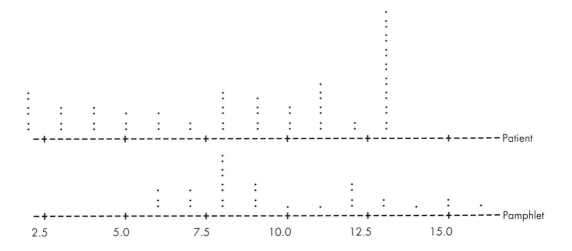

(a) Explain why the form of the data do not allow one to calculate the *mean* reading skill level of patients.

(b) Determine the *median* reading level of patients.

(c) Determine the median readability level of pamphlets.

(d) How do these medians compare? Are they fairly close?

(e) Does the closeness of these medians indicate that the pamphlets are well-matched to the patients' reading levels? Compare the dotplots above to guide your thinking.

(f) What proportion of the patients do not have the reading skill level necessary to read even the simplest pamphlet in the study? (Examine the dotplots above to address this question.)

This activity illustrates that while measures of center are often important, they do not summarize all aspects of a distribution.

Activity 4-5: Students' Distances from Home

Consider the data collected above in the "Preliminaries" section of this Topic, concerning students' distances from home.

(a) Enter the data into the calculator and have the calculator produce a dotplot and a histogram of the distribution.

(b) Would you say that the distribution of distances from home is basically symmetric or skewed in one direction or the other?

(c) Based on the shape of the distribution, do you expect the mean distance from home to be close to the median distance, smaller than the median, or greater than the median?

(d) Have the calculator compute the mean and median distances from home, recording them below. Comment on the accuracy of your expectation in (c).

(e) What is your own distance from home? Is it above or below the mean? How does it compare to the median? About where does your distance from home fall in the distribution?

(f) If you could not look at the dotplot and histogram of these distances but were only told the mean and median distance, would you have a thorough understanding of the distribution of distances from home? Explain.

HOMEWORK ACTIVITIES

Activity 4-6: Planetary Measurements

The following table lists the average distance from the sun (in millions of miles), diameter (in miles), and period of revolution around the sun (in Earth days) for the nine planets of our solar system.

Planet	Distance (million miles)	Diameter (miles)	Revolution (days)
Mercury	36	3,030	88
Venus	67	7,520	225
Earth	93	7,926	365
Mars	142	4,217	687
Jupiter	484	88,838	4,332
Saturn	887	74,896	10,760
Uranus	1,765	31,762	30,684
Neptune	2,791	30,774	60,188
Pluto	3,654	1,428	90,467

(a) Calculate (by hand) the median value of each of these variables.
(b) If a classmate uses the $(n + 1)/2$ formula and obtains a median diameter of 88,838 miles, what do you think would be the most likely cause of his/her mistake? (Notice that this is Jupiter's diameter.)

Activity 4-7: Supreme Court Service *(cont.)*

Use the calculator to display the distribution of years of service for *all* Supreme Court justices who preceded the ones listed in Activity 3-1. These data are listed below:

5	31	32	23	21	21	19	22	6	6	23
1	4	14	6	9	20	3	5	34	13	31
20	34	32	8	14	15	4	8	19	7	3
8	30	28	23	34	10	5	15	23	18	4
6	16	4	18	6	2	26	16	36	7	24
9	18	28	28	7	16	5	7	9	16	17
1	33	15	14	20	13	10	16	5	16	24
13	22	19	34	11	26	10	11	1	33	15
15	20	27	8	5	29	26	15	12	5	14
4	2	5	10							

(a) Describe the general shape of the distribution.
(b) Based on this shape, do you expect the mean to be larger than the median, smaller than the median, or about the same as the median?
(c) Have the calculator compute the mean and median of these years of service. Report these values and comment on your expectation in (b).

(d) How do the mean and median of these years of service compare with those you found in Activity 4-1 for active Supreme Court justices? Offer an explanation for any differences that you find.

Activity 4-8: Students' Measurements *(cont.)*

Consider the data collected in Topic 3, and select *one* of the variables foot length, height, or armspan. Use the calculator to compute the mean and median for this variable. Indicate which variable you are analyzing and report the mean and median. Also comment on whether the mean is greater than, less than, or very close to the median.

Activity 4-9: Tennis Simulations *(cont.)*

Refer back to the data presented in Activity 2-4 concerning lengths (as measured in points) of simulated tennis games.

(a) Determine the mean, median, and mode of the game lengths. (You may use the calculator.)

The experiment also simulated tennis games using the "no-ad" scoring system which awards the game to the first player to reach four points. The data are tallied below:

Points in game	4	5	6	7
Tally (count)	13	22	33	32

(b) Determine the mean, median, and mode of the game lengths with no-ad scoring. Write a sentence or two comparing these with their counterparts from conventional scoring.

Activity 4-10: ATM Withdrawals *(cont.)*

Reconsider the data presented in Activity 3-13 concerning ATM withdrawals. There were a total of 111 withdrawals during the year.

(a) Do the data as presented in that activity enable you to identify the mode value of those 111 withdrawal amounts? If so, identify the mode. If not, explain.

(b) Do the data as presented in that activity enable you to determine the median value of those 111 withdrawal amounts? If so, determine the median. If not, explain.

(c) Do the data as presented in that activity enable you to calculate the mean value of those 111 withdrawal amounts? If so, calculate the mean. If not, explain.

(d) The following table summarizes the individual amounts of the 111 withdrawals. For example, 17 withdrawals were for the amount of $20 and 3 were for the amount of $50; no withdrawals were made of $40. Use this new information to calculate whichever of the mode, median, and mean you could not calculate before.

Amount	$20	$50	$60	$100	$120	$140	$150	$160	$200	$240	$250
Tally	17	3	7	37	3	8	16	10	8	1	1

(e) Create (by hand) a histogram of the distribution of these 111 withdrawals. Comment on key features of the distribution, including its unusual granularity.

Activity 4-11: Gender of Physicians *(cont.)*

Consider the data from Activity 1-5 about the numbers of male and female physicians of different specialties in the U.S.

(a) Use the calculator to produce a dotplot of the distribution of percentage of women in the various specialties. Have the calculator compute the mean and median of this distribution.

(b) Suppose that ten doctors are gathered for a meeting in room A while twenty have convened in room B and fifty in room C. If six of the doctors in room A are women, compared to eight in room B and eleven in room C, calculate the percentage of women in each room.

(c) Find the mean of the three percentages found in (b). Does this equal the overall percentage of women among the doctors in the three rooms combined? Explain.

(d) Does the mean that you calculated in (a) equal the percentage of women among all specialties of U.S. physicians? If not, use the calculator to compute this percentage from the data given.

Activity 4-12: Creating Examples

For each of the following properties, try to construct a data set of ten hypothetical exam scores that satisfies the property. Also produce a dotplot of your hypothetical data in each case. Assume that the exam scores are integers between 0 and 100, inclusive. You may use the calculator.

(a) 90% of the scores are greater than the mean
(b) the mean is greater than twice the mode
(c) the mean is less than two-thirds the median
(d) the mean equals the median but the mode is greater than twice the mean
(e) the mean does not equal the median and none of the scores are between the mean and the median

Activity 4-13: Wrongful Conclusions

For each of the following arguments, explain why the conclusion drawn is not valid. Also include a simple hypothetical example which illustrates that the conclusion drawn need not follow from the information.

(a) A real estate agent notes that the mean housing price for an area is $125,780 and concludes that half of the houses in the area cost more than that.
(b) A businesswoman calculates that the median cost of the five business trips that she took in a month is $600 and concludes that the total cost must have been $3000.
(c) A company executive concludes that an accountant must have made a mistake because she prepared a report stating that 90% of the company's employees earn less than the mean salary.
(d) A restaurant owner decides that more than half of her customers prefer chocolate ice cream because chocolate is the mode when customers are offered their choice of chocolate, vanilla, and strawberry.

WRAP-UP

You have explored in this topic how to calculate a number of measures of the center of a distribution. You have discovered many properties of the *mean*, *median*, and *mode* (such as the important concept of *resistance*) and discovered that these statistics can produce very different values with certain data sets. Most importantly, you have learned that these statistics measure only *one* aspect of a distribution and that you must combine these numerical measures with what you already know about displaying distributions visually and describing them verbally.

In the next topic, you will discover similar measures of another aspect of a distribution of data: its *variability*.

Topic 5:

MEASURES OF SPREAD

OVERVIEW

In the previous topic you explored important numerical measures of the center of a distribution. In this topic you will investigate similar numerical measures of a distribution's variability. These measures will also lead you to discover another visual display (the *boxplot*) and to a very important technique (*standardization*) which will appear throughout the text.

OBJECTIVES

- To learn to calculate certain statistics (*range, interquartile range, standard deviation*) for summarizing the variability of a distribution of data.
- To discover the *five-number summary* of a distribution.
- To explore the *boxplot* as another convenient and informative visual display of a distribution.
- To investigate and determine properties of these summary statistics.
- To understand the *empirical rule* as a means for interpreting the value of standard deviation for certain types of distributions.
- To appreciate the applicability of calculating *z-scores* for comparing distributions of different variables.
- To recognize some of the limitations of these measures of variability.

PRELIMINARIES

1. Take a guess concerning the average high temperature in January in Chicago. Do the same for the average high temperature in San Diego in January.

2. Take a guess concerning the average high temperature in July in Chicago. Do the same for the average high temperature in San Diego in July.

3. Think about the average high temperatures in January for selected cities from across the United States. Would you expect this distribution to be more or less variable than the average high temperatures in July for those same cities?

IN-CLASS ACTIVITIES

Activity 5-1: Supreme Court Service (*cont.*)

Reconsider the data from Activity 4-1 concerning the length of service of Supreme Court Justices.

(a) Create again a dotplot of the distribution of justices' years of service.

(b) Recall from Topic 4 the mean and median of the justices' years of service; record them below.

As we discovered three measures of the center of a distribution, we will encounter three measures of the variability (or spread) of a distribution:

- The *range* is simply the difference between the largest and smallest values in the distribution.

- The *interquartile range* (IQR) is the difference between the lower and upper quartiles (also known as the 25th and 75th percentiles) of the distribution. The *lower quartile* is the value such that (roughly) 25% of the observations fall below it and 75% above it; the *upper quartile* is the value such that (roughly) 75% of the observations fall below it and 25% above it. To find the lower quartile, one finds the median of those observations falling below the location of the actual median; similarly, the upper quartile is the median of those observations falling above the location of the actual median. (When there are an odd number of observations, one does *not* include the actual median in either group.)
- The *standard deviation* is a more complicated measure of variability based on observations' squared deviations from the mean. To compute the standard deviation, one calculates the difference between the mean and each observation and then squares that value for each observation. These squares are then added and the sum is divided by $n - 1$ (one less than the sample size). The standard deviation is the square root of the result.

(c) Calculate the *range* of the distribution of justices' years of service.

(d) To find the lower quartile of the distribution, first list the observations which fall below the location of the median. (Since there are an odd number of observations, do not include the median itself.) Then determine the median of this list.

(e) Similarly, to find the upper quartile of the distribution, list the observations which fall above the location of the median. Then determine the median of this list.

(f) Calculate the interquartile range of the distribution by determining the difference between the quartiles.

(g) To calculate the standard deviation of the distribution, begin by filling in the missing entries in the following table:

	Original data	Deviation from mean	Squared deviation
	22	13.56	183.75
	19	10.56	111.42
	13		
	8	−0.44	0.20
	6		
	4	−4.44	19.75
	3	−5.44	29.64
	1	−7.44	55.42
	0	−8.44	71.31
Column sum	76		

(h) Divide the sum of the last column by 8 (one less than the sample size). Then take the square root of the result to find the standard deviation of the distribution.

Perhaps the introduction of some notation would clarify matters at this point. If we let x_i denote the ith observation and continue to let n denote the number of observations, then we can express the formula for calculating the sample mean (which we will denote by $\bar{x}$) by

$$\bar{x} = \frac{\sum_{i=1}^{n} x_i}{n},$$

where the Σ symbol means to add up what follows it. Similarly, we can express the formula for calculating the sample standard deviation (which we will denote by s or s_x) by

$$s_x = \sqrt{\frac{\sum_{i=1}^{n} (x_i - \bar{x})^2}{n - 1}}.$$

The *five-number summary* of a distribution is composed of the minimum observation, the lower quartile, the median, the upper quartile, and the maximum observation. You met this summary in Topic 3 but have now learned how to calculate these values more precisely. Another graphical display of a distribution known as the *boxplot* is based on this five-number summary. To construct a boxplot, one draws a "box" between the quartiles, thus indicating where the middle 50% of the data fall. Horizontal lines called "whiskers" are then extended from the middle of the sides of the box to the minimum and to the maximum. The median is then marked with a vertical line inside the box.

(i) Use your earlier calculations and the axis drawn below to construct a boxplot of the distribution of justices' years of service. For the sake of comparison, a dotplot has been presented below, but it is not part of the boxplot.

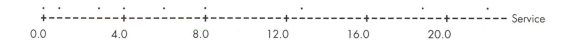

As with measures of center, the abilities to understand the properties of these measures of spread and to interpret them correctly are more important than being able to perform the calculations.

(j) Enter the length of service of Supreme Court Justices from Activity 4-1 into a list named JUSTI.

(k) Set up Plot1 as the following. (Note that there are two icons that look like boxplots. Choose the second one for now. The first icon represents a "modified boxplot" that you will study later):

Now press ZOOM 9 . Compare this boxplot with the one you created by hand.

(l) View the five-number summary by pressing the TRACE button and then using the right and left scroll buttons. On the TI-83, Q_1 denotes the lower quartile and Q_3 denotes the upper quartile.

Activity 5-2: Properties of Measures of Spread

Reconsider the hypothetical exam scores that you first analyzed in Activity 3-1. (The data appear in Appendix A.)

(a) Based on the distributions as revealed in the dotplots in Activity 3-1, how would you expect the standard deviations of exam scores to compare among classes D, E, and F? What about the interquartile ranges among these three classes?

(b) Use the 1-Var Stats command of your calculator to compute standard deviations and inter-quartile ranges for exam scores in classes D, E, and F. The lists,which should already be in your calculator, are named CLSD, CLSE, and CLSF, respectively. Do the calculations support your expectations from (a)?

	Class D	Class E	Class F
Std. dev.			
IQR			

To investigate the effect of outliers on these measures of center, reconsider the data on Supreme Court justices' years of tenure.

(c) Enter the data into the calculator and compute the standard deviation and interquartile range of the years of tenure. Record the results in the first row of the table below.

(d) Now imagine that Justice Rehnquist has served for 42 years rather than 22. Have the calculator recompute the standard deviation and inter-quartile range in this case; record the values in the table.

(e) Finally, suppose that Justice Rehnquist's 22 years of service had been mistakenly recorded as 222 years. Again have the calculator recompute the standard deviation and interquartile range, recording the values in the table.

	Std. dev.	IQR
Justices		
Justices with "big" outlier		
Justices with "huge" outlier		

(f) Based on these calculations, would you say that the standard deviation or the interquartile range is resistant? How about the range? Explain. (Recall the definition of resistance from Topic 3.)

Activity 5-3: Placement Exam Scores

The Department of Mathematics and Calculator Science of Dickinson College gives an exam each fall to freshmen who intend to take calculus; scores on the exam are used to determine into which level of calculus a student should be placed. The exam consists of 20 multiple-choice questions. Scores for the 213 students who took the exam in 1992 are tallied in the following table:

Score	1	2	3	4	5	6	7	8	9	10
Count	1	1	5	7	12	13	16	15	17	32
Score	11	12	13	14	15	16	17	18	19	
Count	17	21	12	16	8	4	7	5	4	

The mean score on this exam is $\bar{x} = 10.221$. The standard deviation of the exam scores is $s = 3.859$. A histogram of this distribution follows:

pl score $N = 213$

Midpoint	Count
2	2
4	12
6	25
8	31
10	49
12	38
14	28
16	12
18	12
20	4

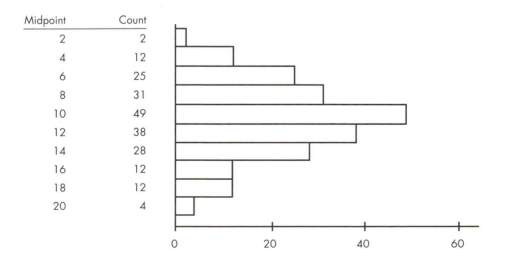

(a) Does this distribution appear to be roughly symmetric and mound-shaped?

(b) Consider the question of how many scores fall within one standard deviation of the mean (denoted by $\bar{x} \pm s$). Determine the upper endpoint of this interval by adding the value of the standard deviation to that of the mean. Then determine the interval's lower endpoint by subtracting the value of the standard deviation from that of the mean.

(c) Look back at the table of tallied scores to determine how many of the 213 scores fall within *one* standard deviation of the mean. What proportion of the 213 scores is this?

(d) Determine how many of the 213 scores fall within *two* standard deviations of the mean, which turns out to be between 2.503 and 17.939. What proportion is this?

(e) Determine how many of the 213 scores fall within *three* standard deviations of the mean, which turns out to be between -1.356 and 21.798. What proportion is this?

You have discovered what is sometimes called the *empirical rule*. It turns out that with mound-shaped distributions, about 68% of the observations fall within one standard deviation of the mean, about 95% fall within two standard deviations of the mean, and virtually all of the observations fall within three standard deviations of the mean. (Notice that this applies to mound-shaped distributions but not necessarily to distributions of other shapes.) This rule provides a way of interpreting the value of a distribution's standard deviation.

Activity 5-4: SATs and ACTs

Suppose that a college admissions office needs to compare scores of students who take the Scholastic Aptitude Test (SAT) with those who take the

American College Test (ACT). Suppose that among the college's applicants who take the SAT, scores have a mean of 896 and a standard deviation of 174. Further suppose that among the college's applicants who take the ACT, scores have a mean of 20.6 and a standard deviation of 5.2.

(a) If applicant Bobby scored 1080 on the SAT, how many points above the SAT mean did he score?

(b) If applicant Kathy scored 28 on the ACT, how many points above the ACT mean did she score?

(c) Is it sensible to conclude that since your answer to (a) is greater than your answer to (b), Bobby outperformed Kathy on the admissions test? Explain.

(d) Determine how many standard deviations above the mean Bobby scored by dividing your answer to (a) by the standard deviation of the SAT scores.

(e) Determine how many standard deviations above the mean Kathy scored by dividing your answer to (b) by the standard deviation of the ACT scores.

This activity illustrates the use of the standard deviation is to make comparisons of individual values from different distributions. One calculates a *z-score* or *standardized score* by subtracting the mean from the value of interest and then dividing by the standard deviation. These *z*-scores indicate how many standard deviations above (or below) the mean a particular value falls. One should only use *z*-scores when working with mound-shaped distributions, however.

(f) Which applicant has the higher *z*-score for his/her admissions test score?

(g) Explain in your own words which applicant performed better on his/her admissions test.

(h) Calculate the z-score for applicant Mike who scored 740 on the SAT and for applicant Karen who scored 19 on the ACT.

(i) Which of the two, Mike or Karen, has the higher z-score?

(j) Under what conditions does a z-score turn out to be negative?

HOMEWORK ACTIVITIES

Activity 5-5: Hypothetical Manufacturing Processes (*cont.*)

Look back at the dotplots from Activity 3-6 of data from hypothetical manufacturing processes. The following table lists the means and standard deviations of these processes. Match each dotplot (process A, B, C, and D) with its numerical statistics (process 1, 2, 3, or 4). (The data appear in Appendix A.)

	Process 1	Process 2	Process 3	Process 4
Mean	12.008	12.004	11.493	11.723
Std. dev.	0.274	0.089	0.041	0.18

Activity 5-6: Climatic Conditions

The following table lists average high temperatures in January and in July for selected cities from across the United States.

City	Jan. hi	July hi	City	Jan. hi	July hi
Atlanta	50.4	88	Nashville	45.9	89.5
Baltimore	40.2	87.2	New Orleans	60.8	90.6
Boston	35.7	81.8	New York	37.6	85.2
Chicago	29	83.7	Philadelphia	37.9	82.6
Cleveland	31.9	82.4	Phoenix	65.9	105.9
Dallas	54.1	96.5	Pittsburgh	33.7	82.6
Denver	43.2	88.2	St. Louis	37.7	89.3
Detroit	30.3	83.3	Salt Lake City	36.4	92.2
Houston	61	92.7	San Diego	65.9	76.2
Kansas City	34.7	88.7	San Francisco	55.6	71.6
Los Angeles	65.7	75.3	Seattle	45	75.2
Miami	75.2	89	Washington	42.3	88.5
Minneapolis	20.7	84			

(a) Calculate (by hand or with the calculator) the interquartile range for the January high temperatures and for the July high temperatures.
(b) Use the calculator to compute the standard deviations for both variables.
(c) Which variable has greater variability in its distribution? Comment on the accuracy of your guess from the "Preliminaries" section.
(d) Which generally has higher temperatures: January or July?
(e) Do you think that if one variable tends to cover larger values than another, then it must have more variability in its values as well? Explain.

Activity 5-7: Planetary Measurements *(cont.)*

Refer back to the data presented in Activity 4-6 concerning planetary measurements.

(a) Calculate (by hand) the five-number summary of distance from the sun.
(b) Draw (by hand) a boxplot of the distribution of distances from the sun.
(c) Would you classify this distribution as roughly symmetric, skewed left, or skewed right?

Activity 5-8: Students' Travels (*cont.*)

Reconsider the data that was collected in Topic 2 concerning the number of *states* visited by students.

(a) Calculate (by hand) the five-number summary of this distribution.
(b) Draw (by hand) a boxplot of this distribution.
(c) Between what two values does the middle 50% of these data fall?

Activity 5-9: Word Lengths *(cont.)*

Reconsider the data collected in Topic 1 concerning the lengths of your words. Calculate (by hand) the five-number summary of this distribution and draw (also by hand) a boxplot. Comment on what the boxplot reveals about the distribution of your word lengths.

Activity 5-10: Tennis Simulations *(cont.)*

Consider the data from Activities 2-4 and 4-9 concerning tennis simulations. Use the calculator to compute the standard deviation of the game lengths with conventional scoring and with no-ad scoring. Write a sentence or two describing your findings.

Activity 5-11: Students' Distances from Home *(cont.)*

Refer back to the data collected in Topic 4 on students' distances from home.

(a) Look at a dotplot of the distribution of distances from home (either hand-drawn or calculator-produced). Based on the shape of this dotplot, would you expect the empirical rule to hold in this case? Explain.
(b) Use the calculator to compute the mean and standard deviation of the distances from home. Report their values.
(c) Determine what proportion of the observations fall within *one* standard deviation of the mean. How closely does this proportion match what the empirical rule would predict?
(d) Determine what proportion of the observations fall within *two* standard deviations of the mean. How closely does this proportion match what the empirical rule would predict?

Activity 5-12: SATs and ACTs *(cont.)*

Refer back to Activity 5-4 in which you calculated *z*-scores to compare test scores of college applicants.

(a) Suppose that applicant Tom scored 820 on the SAT and applicant Mary scored 19 on the ACT. Calculate the *z*-score for Tom and Mary and comment on which of them has the higher test score.

(b) Suppose that scores in a certain state on the Math portion of the SAT have a mean of 474 and a standard deviation of 136, while scores on the Verbal section of the SAT have a mean of 422 and a standard deviation of 122. If Charlie scores 660 on the Math portion and 620 on the Verbal portion, on which portion has he done better in the context of the other test takers in the state?

Activity 5-13: SATs and ACTs *(cont.)*

Refer back to the information about SAT and ACT test scores in Activity 5-4. To answer the following questions, assume that the test scores follow a mound-shaped distribution.

(a) According to the empirical rule, about 95% of SAT takers score between what two values?

(b) What does the empirical rule say about the proportion of students who score between 722 and 1070 on the SAT?

(c) What does the empirical rule say about the proportion of students who score between 10.2 and 31.0 on the ACT?

(d) According to the empirical rule, about 68% of ACT takers score between what two values?

Activity 5-14: Guessing Standard Deviations

Notice that each of the following hypothetical distributions (shown at the top of the next page) is roughly symmetric and mound-shaped.

(a) Use the empirical rule that you discovered above to make an educated guess about the mean and standard deviation of each distribution.

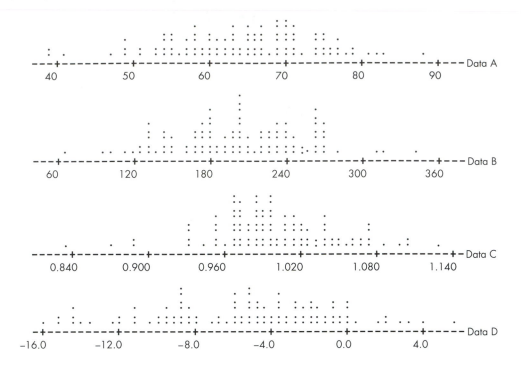

(b) Lists containing the hypothetical distributions data have been stored in a gouped file named HYPOSTDV.83g. Download this grouped file into your calculator. The list names are DATAA, DATAB, DATAC, and DATAD, respectively. Use the calculator to compute the means and standard deviations. Comment on the accuracy of your guesses.

Activity 5-15: Limitations of Boxplots

Consider the hypothetical exam scores presented below for three classes of students. Dotplots of the distributions are also presented.

Class A:	50	50	50	63	70	70	70	71	71	72	72	79	91	91	92
Class B:	50	54	59	63	65	68	69	71	73	74	76	79	83	88	92
Class C:	50	61	62	63	63	64	66	71	77	77	77	79	80	80	92

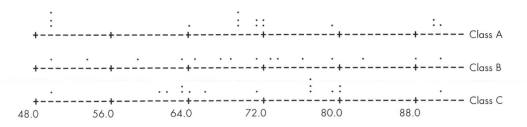

(a) Do these dotplots reveal differences among the three distributions of exam scores? Explain briefly.
(b) Calculate (by hand) five-number summaries of the three distributions. (Notice that the scores are already arranged in order.)
(c) Create boxplots (on the same scale) of these three distributions.
(d) If you had not seen the actual data and had only been shown the box-plots, would you have been able to detect the differences in the three distributions?

Activity 5-16: Creating Examples *(cont.)*

For each of the following properties, try to construct a data set of ten hypothetical exam scores that satisfies the property. Also produce a dotplot of your hypothetical data in each case. Assume that the exam scores are integers between 0 and 100, inclusive. You may use the calculator.

(a) less than half of the exams fall within one standard deviation of the mean
(b) the standard deviation is positive but the interquartile range equals 0
(c) all of the exams fall within one standard deviation of the mean
(d) the standard deviation is as large as possible
(e) the interquartile range is as large as possible but the standard deviation is not as large as possible

Activity 5-17: More Measures of Center

Consider the following three numerical measures of the center of a distribution:

- *trimmed mean:* the mean of the observations after removing the largest 5% and smallest 5%
- *midrange:* the average of the minimum and maximum
- *midhinge:* the average of the lower and upper quartiles

(a) Which of these three measures of center is resistant to outliers and which is not resistant? Explain your answers based on how the measures are defined.
(b) Calculate the midrange and midhinge of the Supreme Court justices' years of service from Activities 4-1 and 5-1. Comment on how these compare to the mean and median found above.

(c) Calculate the midrange and midhinge of the data collected in Topic 2 on states visited by students. Comment on how these compare to the mean and median found in Activity 4-10.

Activity 5-18: Hypothetical ATM Withdrawals

Suppose that a bank wants to monitor the withdrawals that its customers make from automatic teller machines at three locations. Suppose further that they sample 50 withdrawals from each location and tally the data in the following table. All missing entries are zeroes.

Cash amount	$20	$30	$40	$50	$60	$70	$80	$90	$100	$110	$120
Machine 1 tally			25						25		
Machine 2 tally	2	8	1	9	2	6	2	9	1	8	2
Machine 3 tally	9					32					9

(a) Use the calculator to produce visual displays of the distributions of cash amounts at each machine. Is each distribution perfectly symmetric?

(b) Use the calculator to compute the mean and standard deviation of the cash withdrawal amounts at each machine. Are they identical for each machine?

(c) Are the *distributions* themselves identical? Would you conclude that the mean and standard deviation provide a complete summary of a distribution of data?

WRAP-UP

In this topic you have learned to calculate and studied properties of the *range, interquartile range,* and *standard deviation* as measures of the variability of a distribution. You have also discovered a new visual display, the *boxplot,* and studied the *five-number summary* on which it is based. In addition, you have explored the *empirical rule* and *z-scores* as applications of the standard deviation.

To this point you have dealt primarily with one distribution at a time. Often it is more interesting to compare distributions between/among two or more groups. In the next topic, you will discover some new techniques and also apply what you have learned so far to that task.

Topic 6:

COMPARING DISTRIBUTIONS

OVERVIEW

You have been analyzing distributions of data by constructing various graphical displays (dotplot, histogram, stemplot, boxplot), by calculating numerical measures of various aspects of that distribution (mean, median, and mode for center; range, interquartile range, and standard deviation for spread), and by commenting verbally on features of the distribution revealed in those displays and statistics. Thus far you have concentrated on one distribution at a time. With this topic you will apply these techniques in the more interesting case of analyzing, comparing, and contrasting distributions from two or more groups simultaneously.

OBJECTIVES

- To see how to construct *side-by-side stemplots* to compare two distributions of data.
- To understand the meaning of the phrase *"statistical tendency"*.
- To learn a formal rule-of-thumb for determining whether an observation is an *outlier*.
- To discover *modified boxplots* as displays conveying more information than ordinary boxplots.
- To acquire extensive experience with using graphical, numerical, and verbal means of comparing and contrasting distributions from two or more groups.
- To use the calculator as an important tool for performing such analyses.

PRELIMINARIES

1. Which state would you guess was the fastest growing (in terms of population) in the U.S. between 1990 and 1993?

2. Would you expect to find much of a difference in motor vehicle theft rates between eastern and western states? If so, which region would you expect to have higher theft rates?

3. Consider the assertion that "men are taller than women." Does this mean that every man is taller than every woman? If not, write one sentence indicating what you think the assertion does mean.

4. What would you guess for the average lifetime of a notable scientist of the past? How about for a notable writer of the past?

5. Take a guess as to the most money won by a female golfer in 1990. Do the same for a male golfer.

6. Which of the seven seasons of the television show *Star Trek: The Next Generation* would you rate as the best? If you have no opinion, guess which season a panel of reviewers would rate as the best.

7. Take a guess as to how long a typical penny has been in circulation.

8. Of the four common American coins (penny, nickel, dime, quarter), which do you think tends to have the longest lifetime?

9. Pool the spare change of the class and record the years of the first 25 pennies, nickels, dimes, and quarters that you come across:

Pennies	Nickels	Dimes	Quarters

IN-CLASS ACTIVITIES

Activity 6-1: Shifting Populations

The following table lists the percentage change in population between 1990 and 1993 for each of the 50 states.

(a) Indicate in the table whether the state lies mostly to the east (E) or to the west (W) of the Mississippi River.

State	% change	Region	State	% change	Region
Alabama	3.6		Montana	5.1	
Alaska	8.9		Nebraska	1.8	
Arizona	7.4		Nevada	15.6	
Arkansas	3.1		New Hampshire	1.4	
California	3.7		New Jersey	1.9	
Colorado	8.2		New Mexico	6.7	
Connecticut	−0.3		New York	1.1	
Delaware	5.1		North Carolina	4.8	
Florida	5.7		North Dakota	−0.6	
Georgia	6.8		Ohio	2.3	
Hawaii	5.7		Oklahoma	2.7	
Idaho	9.2		Oregon	6.7	
Illinois	2.3		Pennsylvania	1.4	
Indiana	3.0		Rhode Island	−0.3	
Iowa	1.3		South Carolina	4.5	
Kansas	2.1		South Dakota	2.8	
Kentucky	2.8		Tennessee	4.5	
Louisiana	1.9		Texas	6.2	
Maine	0.9		Utah	7.9	
Maryland	3.8		Vermont	2.3	
Massachusetts	−0.1		Virginia	4.9	
Michigan	2.0		Washington	8.0	
Minnesota	3.3		West Virginia	1.5	
Mississippi	2.7		Wisconsin	3.0	
Missouri	2.3		Wyoming	3.7	

In a *side-by-side stemplot*, a common set of stems is used in the middle of the display with leaves for each category branching out in either direction, one to the left and one to the right. The convention is to order the leaves from the middle out toward either side.

(b) Construct a side-by-side stemplot of these population shift percentages according to the state's region; use the stems listed below.

West		East
	|−0|	
	| 0 |	
	| 1 |	
	| 2 |	
	| 3 |	
	| 4 |	
	| 5 |	
	| 6 |	
	| 7 |	
	| 8 |	
	| 9 |	
	| 10 |	
	| 11 |	
	| 12 |	
	| 13 |	
	| 14 |	
	| 15 |	

Remember to arrange the leaves in order from the inside out:

West		East
	|−0|	
	| 0 |	
	| 1 |	
	| 2 |	
	| 3 |	
	| 4 |	
	| 5 |	
	| 6 |	
	| 7 |	
	| 8 |	
	| 9 |	
	| 10 |	
	| 11 |	
	| 12 |	
	| 13 |	
	| 14 |	
	| 15 |	

(c) Calculate the median value of the percentage change in population for each region.

(d) Identify your home state and comment on where it fits into the distribution.

(e) Does one region (east or west) *tend* to have higher percentage changes than the other? Explain.

(f) Is it the case that every state from one region has a higher percentage change than every state from the other? If not, identify a pair such that the eastern state has a higher percentage change than the western state.

(g) If you were to randomly pick one state from each region, which would you expect to have the higher percentage change? Explain.

You have discovered an important (if somewhat obvious) concept in this activity—that of *statistical tendency*. You found that western states *tend* to have higher percentage changes in population than do eastern states. It is certainly not the case, however, that *every* western state has a higher percentage change than *every* eastern state.

Similarly, men *tend* to be taller than women, but there are certainly some women who are taller than most men. Statistical tendencies pertain to average or typical cases but not necessarily to individual cases. Just as Geena Davis and Danny DeVito do not disprove the assertion that men are taller than women, the cases of California and Georgia do not contradict the finding that western states *tend* to have higher percentage changes in population than eastern states.

Activity 6-2: Professional Golfers' Winnings

The following table presents the winnings (in thousands of dollars) of the 30 highest money winners on each of the three professional golf tours (PGA for males, LPGA for females, and Seniors for males over 50 years of age) in 1990. (Even if you are not a golf fan, perhaps you are interested either in making money or in studying earning differences between the genders.)

Rank	PGA golfer	Win-nings	LPGA golfer	Win-nings	Seniors golfer	Win-nings
1	Norman	1165	Daniel	863	Trevino	1190
2	Levi	1024	Sheehan	732	Hill	895
3	Stewart	976	King	543	Coody	762
4	Azinger	944	Gerring	487	Archer	749
5	Mudd	911	Bradley	480	Rodriguez	729
6	Irwin	838	Jones	353	Dent	693
7	Calcavecchia	834	Okamoto	302	Charles	584
8	Simpson	809	Lopez	301	Douglass	568
9	Couples	757	Ammacapane	300	Player	507
10	O'Meara	707	Rarick	259	McBee	480
11	Morgan	702	Coe	240	Crampton	464
12	Mayfair	693	Mochrie	231	Henning	409
13	Wadkins	673	Walker	225	Geiberger	373
14	Mize	668	Johnson	187	Hill	354
15	Kite	658	Richard	186	Nicklaus	340
16	Baker-Finch	611	Geddes	181	Beard	327
17	Beck	571	Keggi	180	Mowry	314
18	Elkington	548	Crosby	169	Thompson	308
19	Jacobsen	547	Massey	166	Dill	278
20	Love	537	Figg-Currier	157	Zembriski	276
21	Grady	527	Johnston	156	Barber	274
22	Price	520	Green	155	Moody	273
23	Tway	495	Mucha	149	Bies	265
24	Roberts	478	Eggeling	147	Kelley	263
25	Gallagher	476	Rizzo	145	Jimenez	246
26	Pavin	468	Brown	140	Shaw	235
27	Gamez	461	Mallon	129	Massengale	229
28	Cook	448	Hammel	128	January	216
29	Tennyson	443	Benz	128	Cain	208
30	Huston	435	Turner	122	Powell	208

(a) Notice that one cannot construct side-by-side stemplots to compare these distributions since there are three groups and not just two to be compared. One can, however, construct comparative boxplots of the distributions. Start by calculating (by hand) five-number summaries for each of the three groups, recording them below. Notice that the observations have already been arranged in order and that they are numbered, which should greatly simplify your calculations.

Tour	Minimum	Lower quartile	Median	Upper quartile	Maximum
PGA					
LPGA					
Senior					

(b) Construct (by hand) boxplots of these three distributions on the same scale; the axis has been drawn for you below.

(c) The boxplot for the Senior golfers reveals one of the weaknesses of boxplots as visual displays. Clearly Lee Trevino, the top money winner among the Seniors, was an outlier, earning almost $300,000 more than his nearest competitor. Would the boxplot look any different, however, if his nearest competitor had won only $5 less than him?

One way to address this problem with boxplots is to construct what are sometimes called *modified boxplots*. These treat outliers differently by marking them with a special symbol (*) and then only extending the boxplot's "whiskers" to the most extreme nonoutlying value. To do this requires an explicit rule for identifying outliers. The rule that we will use regards outliers as observations lying more than 1.5 times the interquartile range away from the nearer quartile.

With the Seniors' winnings the lower quartile is $Q_1 = 265$ and the upper quartile is $Q_3 = 568$, so the inter-quartile range is IQR $= 568 - 265 = 303$. Thus, $1.5 \times$ IQR $= 1.5 (303) = 454.5$, so any observation more than 454.5 away from its nearer quartile will be considered an outlier. To look for such observations, we add 454.5 to Q_3, obtaining $568 + 454.5 = 1,022.5$. Since Trevino's 1,190 exceeds this, it is an outlier. On the other end, we subtract 454.5 from Q_1; since the result is negative, clearly no observations fall below it, so there are no outliers on the low end.

(d) Use this rule to check for and identify outliers on the PGA and LPGA lists.

Modified boxplots are constructed by marking outliers with an "*" and extending "whiskers" only to the most extreme (i.e., largest on the high end, smallest on the low end) *non*outlier.

(e) Construct modified boxplots of the three distributions below.

(f) Lists containing the profssional golfer's winnings have been stored in a grouped file named GOLFWINN.83g. Download this grouped file into your calculator (the lists are named PGA, LPGA, and SEN, respectively) and create side-by-side modified boxplots on your calculator by using Plot1, Plot2, and Plot3 simultaneously. For example, you might set up Plot1 as follows:

Compare the calculator-generated modified boxplots with your hand-drawn modified boxplots above.

(g) Do the modified boxplots provide more visual information than the "un-modified" ones? Explain.

(h) Write a paragraph comparing and contrasting key features of the distributions of the earnings of the top money winners on these three golf tours.

Activity 6-3: Ages of Coins

Consider the data collected in class regarding ages of coins. Enter the ages into the calculator. (You might want to enter the dates and then use the calculator to compute the ages.)

(a) Use the calculator to compute the following summary statistics for each coin's distribution of ages.

	Mean	Std. dev.	Mini- mum	Lower quartile	Median	Upper quartile	Maxi- mum
Pennies							
Nickels							
Dimes							
Quarters							

(b) Use these five-number summaries to construct boxplots (on the same scale) of the age distributions. (You need not use modified boxplots.)

(c) Do the data suggest that some coins tend to have longer or shorter circulation periods than others? Explain.

(d) Do the data suggest any dramatic differences in the distributions of coin's ages? If so, describe them.

HOMEWORK ACTIVITIES

Activity 6-4: Students' Measurements *(cont.)*

Refer back to the data collected in Topic 3 and select *one* of the three variables: foot length, height, or armspan. Compare men's and women's distributions of this variable by answering the following questions, supplying displays and/or calculations to support your answers:

(a) Does one gender *tend* to have larger values of this variable than the other gender? If so, by about how much do the genders differ on the average with respect to this variable? Does *every* member of one gender have a larger value of this variable than *every* member of the other gender?

(b) Does one gender have more variability in the distribution of these measurements than the other gender?

(c) Are the shapes of both gender's distributions fairly similar?

(d) Does either gender have outliers in their distribution of this variable?

Activity 6-5: Students' Travels *(cont.)*

Reconsider the data collected in Topic 2 concerning the number of states visited by college students. Construct (by hand) a side-by-side stemplot which compares the distributions of men and women. Then write a paragraph comparing and contrasting the distributions.

Activity 6-6: Tennis Simulations *(cont.)*

Reconsider the data from Activities 2-4 and 4-9 concerning tennis simulations. The experiment also analyzed a "handicap" scoring method which uses no-ad scoring and also awards weaker players bonus points at the start of a game. The lengths of 100 simulated games with this handicap scoring system are tallied below:

Points in game	1	2	3	4	5	6	7
Tally (count)	3	4	12	18	28	25	10

(a) Use the calculator to compute means and standard deviations of the game lengths for each of the three scoring systems.

(b) Write a paragraph addressing whether one method tends to produce longer or shorter games than the others and whether one method produces more or less variability in game lengths than the others.

Activity 6-7: Automobile Theft Rates

Investigate whether states in the eastern or western part of the United States tend to have higher rates of motor vehicle thefts. The following table divides states according to whether they lie east or west of the Mississippi River and lists their 1990 rate of automobile thefts per 100,000 residents.

Eastern state	Theft rate	Western state	Theft rate
Alabama	348	Alaska	565
Connecticut	731	Arizona	863
Delaware	444	Arkansas	289
Florida	826	California	1016
Georgia	674	Colorado	428
Illinois	643	Hawaii	381
Indiana	439	Idaho	165
Kentucky	199	Iowa	170
Maine	177	Kansas	335
Maryland	709	Louisiana	602
Massachusetts	924	Minnesota	366
Michigan	714	Missouri	539
Mississippi	208	Montana	243
New Hampshire	244	Nebraska	178
New Jersey	940	Nevada	593
New York	1043	New Mexico	337
North Carolina	284	North Dakota	133
Ohio	491	Oklahoma	602
Pennsylvania	506	Oregon	459
Rhode Island	954	South Dakota	110
South Carolina	386	Texas	909
Tennessee	572	Utah	238
Vermont	208	Washington	447
Virginia	327	Wyoming	149
West Virginia	154		
Wisconsin	416		

(a) Create a side-by-side stemplot to compare these distributions. Ignore the last digit of each state's rate; use the hundreds digit as the stem and the tens digit as the leaf.

(b) Calculate (by hand) the five-number summary for each distribution of automobile theft rates.

(c) Conduct (by hand) the outlier test for each distribution. If you find any outliers, identify them (by state).

(d) Construct (by hand) modified boxplots to compare the two distributions.

(e) Write a paragraph describing your findings about whether motor vehicle theft rates tend to differ between eastern and western states.

Activity 6-8: Lifetimes of Notables

The 1991 World Almanac and Book of Facts contains a section in which it lists "noted personalities." These are arranged according to a number of categories, such as "noted writers of the past" and "noted scientists of the past." One can calculate (approximately, anyway) the lifetimes of these people by subtracting their year of birth from their year of death. Distributions of the lifetimes of the people listed in nine different categories have been displayed in the following boxplots.

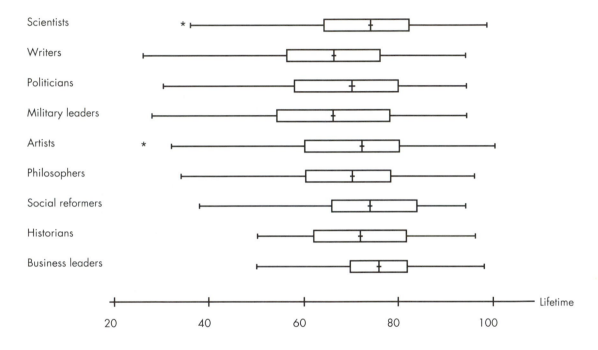

Use the information contained in these boxplots to answer the following questions. (In cases where the boxplots are not clear enough to make a definitive determination, you will have to make educated guesses.)

(a) Which group has the *individual* with the longest lifetime of anyone listed; about how many years did he/she live?

(b) Which group has the largest *median* lifetime; what (approximately) is that median value?

(c) Which group has the largest *range* of lifetimes; what (approximately) is the value of that range?

(d) Which group has the largest *interquartile range* of lifetimes; what (approximately) is the value of that IQR?

(e) Which group has the smallest *interquartile range* of lifetimes; what (approximately) is the value of that IQR?

(f) Which group has the smallest *median* lifetime; what (approximately) is the value of that median?

(g) Describe the general shape of the distributions of lifetimes.

(h) Suggest an explanation for writers tending to live shorter lives than scientists.

Activity 6-9: Hitchcock Films *(cont.)*

Reconsider the data from Activity 3-8 concerning running times of movies directed by Alfred Hitchcock. Perform (by hand) the outlier test to determine if any of the films constitute outliers in terms of their running times. Comment on your findings in light of your analysis in Activity 3-8.

Activity 6-10: Value of Statistics *(cont.)*

Compare the responses of men and women to the question from Topic 1 about rating the value of statistics in society. Write a paragraph or two describing and explaining your findings; include whatever displays or calculations you care to.

Activity 6-11: Governors' Salaries

The following boxplots display the distributions of the 1993 governors' salaries according to the state's geographic region of the country (the * and 0 represent outliers). Region 1 is the Northeast, 2 the Midwest, 3 the South, and 4 the West. Write a paragraph comparing and contrasting key features of these distributions.

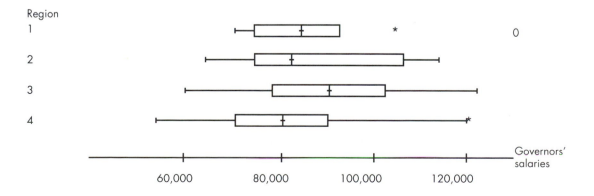

Activity 6-12: Voting for Perot *(cont.)*

Reconsider the data from Activity 3-14 concerning the percentage of votes received by Ross Perot in each state (and D.C.) in the 1992 Presidential election. Use the calculator to analyze possible differences in the vote percentage distributions among the four regions of the country. Write a paragraph summarizing your findings, paying particular attention to the question of whether Perot tended to receive more support in certain regions and less in others.

Activity 6-13: Cars' Fuel Efficiency

Consumer Reports magazine puts out a special issue with information about new cars. The following boxplots display the distributions of cars' miles-per-gallon ratings. The category designations are those of the magazine: 1=small, 2=compact, 3=sports, 4=midsized, 5=large, 6=luxury.

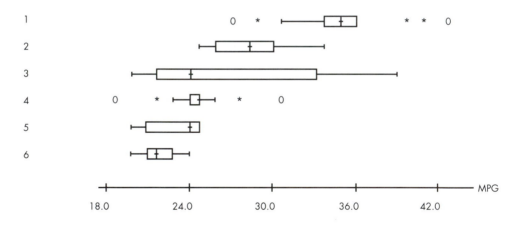

(a) Which category of car tends to get the best fuel efficiency? second best? worst?
(b) Does every car in the best fuel efficiency group have a higher miles-per-gallon rating than every car in the second best group?
(c) Does every car in the best fuel efficiency group have a higher miles-per-gallon rating than every car in the worst fuel efficiency group?
(d) Which group has the most variability in miles-per-gallon ratings?

Activity 6-14: Word Lengths (*cont.*)

The following table lists the lengths of the 26 words used in the first sentence of the "Overview" section of Topic 1:

10	2	3	7	2	9	4	4	2	1	7	5	2
5	4	5	2	2	9	4	2	5	2	3	4	4

(a) Compare the distribution of *my* word lengths with the data that you collected in Topic 1 on the lengths of *your* words. Write a brief paragraph describing your findings; include any visual displays and numerical calculations that you find informative.

(b) Select a sentence or two by a favorite writer of yours and analyze her/his word length distribution in comparison with yours and mine. Again write a brief paragraph describing your findings.

Activity 6-15: Mutual Fund Returns

Mutual funds are collections of stocks and bonds which have become a very popular investment option in recent years. Some funds require the buyer to pay a fee called a "load" which is similar to a commission charge. A natural question to ask is whether funds that require a load charge perform better than funds which require no load. The table below lists the 1993 percentage return for the mutual funds of the Fidelity Investment group:

No load					Load	
13.9	17.8	21.9	12.4	22.1	24.5	19.5
15.4	13.9	12.5	9	9.1	33.4	20.2
23.3	36.7	13.1	13.1	5.2	21.4	24.7
26.3	81.8	13.9	6.7	13.1	26.4	24.7
19.3	21.3	11.2	13.2	14	26.8	8.3
13.8	18.9	35.1	12.9	19.1	19.9	40.1
13.6	9.8	16.2	12.8	10.2	27.2	63.9
25.5	6.5	20.5	12.6	15.6	16.2	21.1
25.9	18.4	12.2	21.4	22.9		
8.1	5.5	13.8	12.5	36.5		

Analyze these data to address the question of whether funds that charge load fees tend to perform better than those without loads. Write a brief summary of your findings, including whatever visual displays or summary statistics you deem appropriate. Also be sure to comment on what it means to say that one type of fund *tends* to outperform another. (You may use the calculator.)

Activity 6-16: *Star Trek* Episodes

Editors of an *Entertainment Weekly* publication ranked every episode of *Star Trek: The Next Generation* from best (rank 1) to worst (rank 178), as shown in the table, separated according to the season of the show's seven-year run in which the episode aired.

Season 1	Season 2	Season 3	Season 4	Season 5	Season 6	Season 7
50	103	113	8	5	23	146
51	90	143	153	16	68	157
101	104	151	25	109	57	78
164	70	91	77	140	127	121
173	105	92	120	76	81	97
148	62	95	108	6	18	61
163	48	160	38	20	3	168
149	161	67	69	42	13	80
175	31	100	112	135	122	128
40	4	87	167	60	64	83
84	172	33	118	39	36	14
139	94	123	82	73	116	58
125	176	12	166	53	117	141
34	75	35	115	147	59	177
162	132	37	144	47	43	124
145	156	88	129	26	49	110
86	52	98	65	7	29	56
155	66	133	137	71	107	134
85	165	72	46	138	32	54
63	99	119	171	111	158	136
150	106	19	21	152	55	74
169	45	89	30	24	11	93
44	178	102	28	79		174
41	126	9	96	15		114
142	159		170	130		1
			17	131		2
			22	27		
				154		
				10		

Five-number summaries for these rankings by season appear below along with comparative boxplots:

	No.	Minimum	Lower quartile	Median	Upper quartile	Maximum
Season 1	25	34	57	139	158.5	175
Season 2	25	4	64	103	157.5	178
Season 3	24	9	44.5	91.5	117.5	160
Season 4	27	8	30	96	137	171
Season 5	29	5	22	60	130.5	154
Season 6	22	3	27.5	56	109.25	158
Season 7	26	1	60.25	103.5	137.25	177

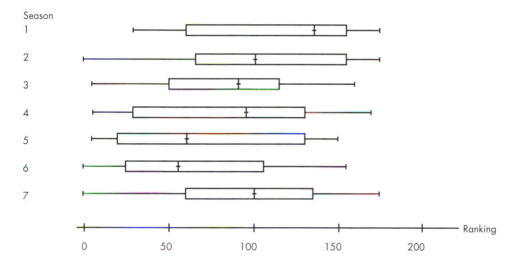

(a) In which season did the highest ranked episode appear?

(b) In which season did the lowest ranked episode appear?

(c) Do any seasons not have a single episode in the top ten? If so, identify it/them.

(d) Do any seasons not have a single episode in the bottom ten? If so, identify it/them.

(e) Which season seemed to have the best episodes according to these reviewers? Explain your choice.

(f) Which season seemed to have the worst episodes according to these reviewers? Explain your choice.

(g) Comment on whether the five-number summaries and boxplots reveal any tendencies toward episodes getting better or worse as the seasons progressed (again, of course, accepting the ranking judgments of these reviewers).

Activity 6-17: Shifting Populations *(cont.)*

Refer back to the data in Activity 6-1 concerning states' population shifts.

(a) Identify the eastern states with the biggest and the smallest percentage increases in population. Do the same for the western states. Also record the values of those percentage increases.

(b) Calculate (by hand) five-number summaries of the percentage changes in population for eastern states and for western states.

(c) Perform (by hand) the outlier test for both regions, identifying any outliers that you find.

(d) Construct (by hand) comparative modified boxplots of these distributions.

(e) Comment on what these boxplots reveal about the distributions of percentage changes in population between eastern and western states.

WRAP-UP

You have been introduced in this topic to methods of *comparing* distributions between/among two or more groups. This task has led you to the very important concept of statistical *tendencies*. You have also expanded your knowledge of visual displays of distributions by encountering *side-by-side stemplots* and *modified boxplots*. Another object of your study has been a formal test for determining whether an unusual observation is an outlier.

In the next unit you will begin to study the important issue of exploring *relationships* between variables.

Unit Two

Exploring Data: Relationships

Topic 7:

GRAPHICAL DISPLAYS OF ASSOCIATION

OVERVIEW

To this point you have been investigating and analyzing distributions of a single variable. This topic introduces you to another topic that plays an important role in statistics: exploring relationships between variables. You will investigate the concept of *association* and explore the use of graphical displays (namely, *scatterplots*) as you begin to study relationships between variables.

OBJECTIVES

- To understand the concept of *association* between two variables and the notions of the *direction* and *strength* of the association.
- To learn to construct and to interpret *scatterplots* as graphical displays of the relationship between two variables.
- To discover the utility of including a 45° line on a scatterplot when working with *paired data*.
- To become familiar with *labeled scatterplots* as devices for including information from a categorical variable into a scatterplot.
- To use the calculator to explore associations between variables of genuine data in a variety of applications.

PRELIMINARIES

1. Do you expect that there is a tendency for heavier cars to get worse fuel efficiency (as measured in miles-per-gallon) than lighter cars?

2. Do you think that if one car is heavier than another that it must always be the case that it gets a worse fuel efficiency?

3. Take a guess as to a typical temperature for a space shuttle launch in Florida.

4. Do you think that people from large families tend to marry people from large families, people from small families, or do you think that there is no relationship between the family sizes of married couples?

5. Do you think that people from large families tend to have large families of their own, small families of their own, or do you think that there's no relationship there?

6. Record in the following table the number of siblings (brothers and sisters) of each student in the class, the number of siblings of each student's father, and the number of siblings of each student's mother. So that everyone "counts" the same way, decide as a class whether or not to count step-siblings and half-siblings.

Stu-dent	Own siblings	Father's siblings	Mother's siblings	Stu-dent	Own siblings	Father's siblings	Mother's siblings
1				13			
2				14			
3				15			
4				16			
5				17			
6				18			
7				19			
8				20			
9				21			
10				22			
11				23			
12				24			

IN-CLASS ACTIVITIES

Activity 7-1: Cars' Fuel Efficiency (*cont.*)

Refer back to Activity 6-13, which dealt with a *Consumer Reports* issue examining 1995 cars. For a small group of 16 car models, the following table lists the weight of the car (in pounds) and the fuel efficiency (in miles-per-gallon) achieved in a 150-mile test drive.

Model	Weight	MPG	Model	Weight	MPG
BMW 3-Series	3250	28	Ford Probe	2900	28
BMW 5-Series	3675	23	Ford Taurus	3345	25
Cadillac Eldorado	3840	19	Ford Taurus SHO	3545	24
Cadillac Seville	3935	20	Honda Accord	3050	31
Ford Aspire	2140	43	Honda Civic	2540	34
Ford Crown Victoria	4010	22	Honda Civic del Sol	2410	36
Ford Escort	2565	34	Honda Prelude	2865	30
Ford Mustang	3450	22	Lincoln Mark VIII	3810	22

The simplest graphical means of displaying two measurement variables simultaneously is a *scatterplot,* which uses a vertical axis for one of the variables and a horizontal axis for the other and places a dot for each observation pair at the intersection of its two values. If you are interested in using the value of one variable to predict the value of another variable, the convention is to place the variable to be predicted (known as the *depen-*

dent variable) on the vertical axis and the variable to do the predicting (the *independent variable*) on the horizontal axis.

(a) Use the axes below to construct a scatterplot of miles-per-gallon vs. weight.

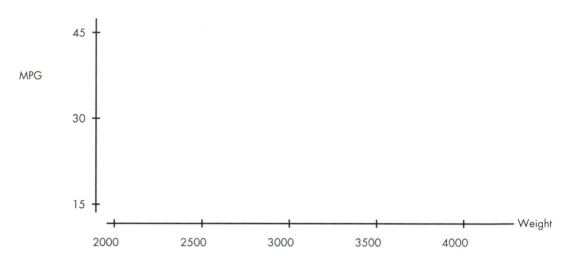

(b) Does the scatterplot reveal any relationship between a car's weight and its fuel efficiency? In other words, does knowing the weight reveal any information at all about the fuel efficiency? Write a few sentences about the relationship between the two variables.

Two variables are said to be *positively associated* if larger values of one variable tend to occur with larger values of the other variable; they are said to be *negatively associated* if larger values of one variable tend to occur with smaller values of the other. The strength of the association depends on how closely the observations follow that relationship. In other words, the strength of the association reflects how accurately one could predict the value of one variable based on the value of the other variable.

(c) Is fuel efficiency positively or negatively associated with weight?

(d) Can you find an example where one car weighs more than another and still manages to have a better fuel efficiency than that other car? If so, identify such a pair and circle them on the scatterplot above.

Clearly the concept of *association* is an example of a statistical *tendency*. It is *not always* the case that a heavier car is less fuel efficient, but heavier cars certainly do *tend* to be.

Activity 7-2: Guess the Association

Consider the following scatterplot of hypothetical scores on the first and second exams of a course.

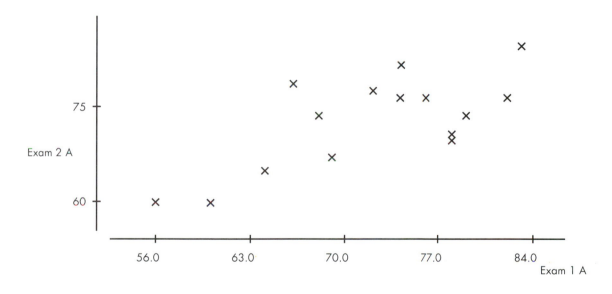

(a) Describe briefly (in words related to the context) what the scatterplot reveals about the relationship between scores on the first exam and scores on the second exam. In other words, does knowing a student's score on the first exam give you any information about his/her score on the second exam? Explain.

(b) Below you will find five more scatterplots of hypothetical exam scores. Your task is to evaluate the direction and the strength of the association between scores on the first exam and scores on the second exam for each of the hypothetical classes A–F. Do so by filling in the table below with the letter (A–F) of the class whose exam scores have the indicated direction and strength of association. (You are to use each class letter once and only once, so you might want to look through all six before assigning letters.)

	Most strong	Moderate	Least strong
Negative			
Positive			

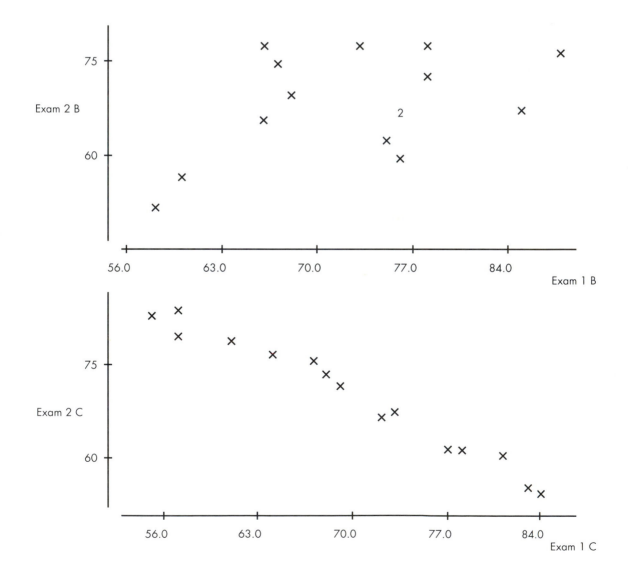

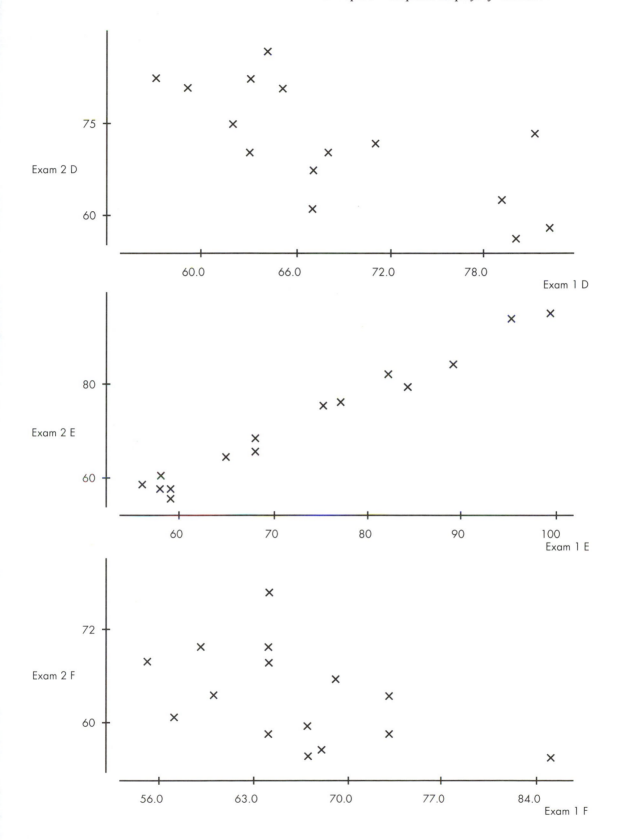

(c) Indicate what you would expect for the direction (positive, negative, or none at all) and strength (none, weak, moderate, or strong) of the association between the pairs of variables listed below.

Pair of variables	Direction of association	Strength of association
Height and armspan		
Height and shoe size		
Height and G.P.A.		
SAT score and college G.P.A.		
Latitude and avg. January temp. of American cities		
Lifespan and weekly cigarette consumption		
Serving size and calories of fast food sandwiches		
Air fare and distance to destination		
Cost and quality rating of peanut butter brands		
Governor's salary and avg. pay in the state		
Course enrollment and average student evaluation		

Activity 7-3: Marriage Ages (*cont.*)

Refer to the data from Activity 3-7 concerning the ages of 24 couples applying for marriage licenses. The following scatterplot displays the relationship between husband's age and wife's age. The line drawn on the scatterplot is a 45° line where the husband's age would equal the wife's age.

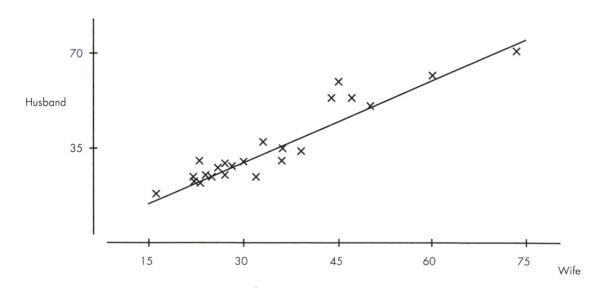

(a) Does there seem to be an association between husband's age and wife's age? If so, is it positive or negative? Would you characterize it as strong, moderate, or weak? Explain.

(b) Look back at the original listing of the data to determine how many of the 24 couples' ages fall exactly on the line. In other words, how many couples listed the same age for both the man and the woman on their marriage license?

(c) Again looking back at the data, for how many couples is the husband older than the wife? Do these couples fall above or below the line drawn in the scatterplot?

(d) For how many couples is the husband younger than the wife? Do these couples fall above or below the line drawn in the scatterplot?

(e) Summarize what one can learn about the ages of marrying couples by noting that the majority of couples produce points which fall above the 45° line.

This activity illustrates that when working with paired data, including a 45° line on a scatterplot can provide valuable information about whether one member tends to have a larger value of the variable than the other member of the pair.

A *categorical* variable can be incorporated into a scatterplot by constructing a *labeled scatterplot*, which assigns different labels to the dots in a scatterplot. For example, one might use an "x" to denote a man's exam score and an "o" to denote a woman's.

Activity 7-4: Fast Food Sandwiches

The following table lists some nutritional information about sandwiches of-fered by Arby's fast food chain. Serving sizes are in ounces.

Sandwich	Meat	Serving	Calories
Regular Roast Beef	Roast beef	5.5	383
Beef and Cheddar	Roast beef	6.9	508
Junior Roast Beef	Roast beef	3.1	233
Super Roast Beef	Roast beef	9	552
Giant Roast Beef	Roast beef	8.5	544
Chicken Breast Fillet	Chicken	7.2	445
Grilled Chicken Deluxe	Chicken	8.1	430
French Dip	Roast beef	6.9	467
Italian Sub	Roast beef	10.1	660
Roast Beef Sub	Roast beef	10.8	672
Turkey Sub	Turkey	9.7	533
Light Roast Beef Deluxe	Roast beef	6.4	294
Light Roast Turkey Deluxe	Turkey	6.8	260
Light Roast Chicken Deluxe	Chicken	6.8	276

(a) Create (by hand) a labeled scatterplot of calories vs. serving ounces, us-ing different labels for the three types of meat (roast beef, chicken, turkey). For example, you might use "r" for roast beef, "c" for chicken, and "t" for turkey.

(b) Disregarding for the moment the distinction between the types of meat, does the scatterplot reveal an association between a sandwich's serving size and its calories? Explain.

(c) Now comment on tendencies you might observe with regard to the type of meat. Does one type of meat tend to have more or fewer calories than other types of similar serving size? Elaborate.

Activity 7-5: Space Shuttle O-Ring Failures

The following data were obtained from 23 shuttle launches prior to *Challenger*'s fatal launch. Each of four joints in the shuttle's solid-fuel rocket motor is sealed by two O-ring joints. After each launch, the reusable rocket motors were recovered from the ocean. This table lists the number of O-ring seals showing evidence of thermal distress and the launch temperature for each of the 23 flights.

Flight date	O-ring failures	Temperature	Flight date	O-ring failures	Temperature
4/12/81	0	66	11/8/84	0	67
11/12/81	1	70	1/24/85	3	53
3/22/82	0	69	4/12/85	0	67
11/11/82	0	68	4/29/85	0	75
4/4/83	0	67	6/17/85	0	70
6/18/83	0	72	7/29/85	0	81
8/30/83	0	73	8/27/85	0	76
11/28/83	0	70	10/3/85	0	79
2/3/84	1	57	10/30/85	2	75
4/6/84	1	63	11/26/85	0	76
8/30/84	1	70	1/12/86	1	58
10/5/84	0	78			

(a) Lists containing the information in the table above have been stored in a grouped file named CHALLENG.83g. Download this grouped file into your calculator. The lists have been named ORING and TEMP, respectively. Use the calculator to construct a scatterplot of O-ring failures vs. temperature. To create a scatterplot of O-ring vs. temperature you need to set up Plot1 as:

The "mark" prompt offers you three choices for the style of point to be displayed. Make sure all other graphing functions are turned off and then press ┌─────┐┌─────┐
 │ZOOM││ 9 │.
 └─────┘└─────┘

(b) Write a few sentences commenting on whether the scatterplot reveals any association between 0-ring failures and temperature.

(c) The forecasted low temperature for the morning of the fateful launch was 31°F. What does the scatterplot reveal about the likelihood of O-ring failure at such a temperature?

(d) Eliminate for the moment those flights which had no O-ring failures. Considering just the remaining seven flights, ask the calculator to construct a new scatterplot of O-ring failures vs. temperature. Does *this* scatterplot reveal association between O-ring failures and temperature? If so, does it make the case as strongly as the previous scatterplot did?

(e) NASA officials argued that flights on which no failures occurred provided no information about the issue. They therefore examined only the second scatterplot and concluded that there was little evidence of an association between temperature and O-ring failures. Comment on the wisdom of this approach and specifically on the claim that flights on which no failures occurred provided no information.

HOMEWORK ACTIVITIES

Activity 7-6: Students' Family Sizes

Enter the data that you collected above concerning family sizes into the calculator.

(a) First examine just the distribution of *students'* siblings by itself. Comment briefly on the distribution, and also report the mean, standard deviation, and five-number summary.

(b) Now examine the *relationships* between students' siblings and fathers' siblings, between students' siblings and mothers' siblings, and between fathers' siblings and mothers' siblings. Have the calculator produce the relevant scatterplots. Write a paragraph summarizing your findings concerning whether an association exists between any pairs of these variables.

Activity 7-7: Air Fares

The table below lists distances (in miles) and cheapest airline fares (in dollars) to certain destinations for passengers flying out of Baltimore, Maryland (as of January 8, 1995):

Destination	Distance	Airfare	Destination	Distance	Airfare
Atlanta	576	178	Miami	946	198
Boston	370	138	New Orleans	998	188
Chicago	612	94	New York	189	98
Dallas/Fort Worth	1216	278	Orlando	787	179
Detroit	409	158	Pittsburgh	210	138
Denver	1502	258	St. Louis	737	98

(a) Create (by hand) a scatterplot of airfare vs. distance.

(b) Is airfare associated with distance? If so, is the association positive or negative? Would you characterize it as strong, moderate, or weak?

(c) Find at least two examples of pairs of cities that go against the association. In other words, if the association is positive, find two pairs of cities where the more distant city is cheaper to fly to. Circle these pairs of cities on your scatterplot.

(d) Explain in your own words how (b) and (c) address the issue of association as a statistical *tendency*.

Activity 7-8: Broadway Shows *(cont.)*

Reconsider the data from Activity 2-7 concerning Broadway shows. Notice that there are two different types of shows listed—plays and musicals.

(a) Produce (by hand) a labeled scatterplot of receipts vs. *percentage* capacity using different labels to distinguish plays from musicals.
(b) Ignoring for the moment the distinction between plays and musicals, comment on the association between receipts and percentage capacity.
(c) Now comment on differences in the relationship between receipts and percentage capacity based on whether the show is a play or a musical. For shows with similar percentage capacities, does one type of show tend to make more money than the other type? Explain.

Activity 7-9: Students' Measurements *(cont.)*

Reconsider the data collected in Topic 3 concerning students' heights and armspans.

(a) Use the calculator to create a scatterplot of height vs. armspan.
(b) Comment on the association between height and armspan as revealed in the scatterplot.
(c) Can you find examples of pairs of students in which the shorter student has the longer armspan?
(d) Draw a 45° line on the scatterplot. The equation for the 45° line is $y = x$. To draw this line on your calculator press $\boxed{\text{Y=}}$, clear all functions, and then enter an "X" at the "Y1=" prompt. Now graph.
(e) What can you say about students who fall *above* the 45° line?
(f) What is true about the ratio of height to armspan for those students who fall *below* the 45° line?

Activity 7-10: Students' Measurements *(cont.)*

Reconsider the data collected in Topic 3 concerning students' heights and foot lengths.

(a) Enter the data concerning the students' heights, footlengths, and gender into your calculator. Since your calculator accepts only numeric values in lists, you need to use a zero to represent male and a one to represent female.
(b) In order for your calculator to produce a labeled scatterplot you need to use the program LBLSCAT.83p. Download this program and then use the calculator to create a labeled scatterplot of height vs. foot length,

using gender as the categorical variable, and male and female as the label 1 and label 2 names.

(c) Disregarding gender for the moment, write a sentence or two to describe the association between height and foot length.

(d) Comment on any gender differences that the scatterplot reveals.

Activity 7-11: College Alumni Donations

The following table lists the number of alumni on the roll for each of Dickinson College's graduating classes and also the number who made a financial contribution to the college between July 1, 1991 and June 30, 1992.

Class	Roll	Donors	Class	Roll	Donors	Class	Roll	Donors
1919	14	2	1944	77	40	1969	348	170
1920	10	2	1945	49	28	1970	354	146
1921	10	2	1946	65	33	1971	389	175
1922	9	2	1947	84	43	1972	371	174
1923	15	3	1948	162	69	1973	446	203
1924	30	12	1949	174	64	1974	397	168
1925	20	6	1950	223	89	1975	426	171
1926	46	11	1951	172	74	1976	453	171
1927	37	7	1952	172	88	1977	442	177
1928	37	22	1953	147	60	1978	401	173
1929	38	21	1954	173	79	1979	390	134
1930	46	28	1955	170	71	1980	435	155
1931	69	29	1956	194	92	1981	494	171
1932	60	29	1957	177	110	1982	457	144
1933	62	35	1958	172	89	1983	461	168
1934	70	40	1959	200	89	1984	444	134
1935	85	45	1960	236	100	1985	526	149
1936	83	43	1961	234	108	1986	500	139
1937	70	33	1962	232	87	1987	522	170
1938	76	35	1963	283	138	1988	502	130
1939	87	44	1964	267	126	1989	553	146
1940	83	39	1965	255	95	1990	577	139
1941	92	44	1966	287	116	1991	609	179
1942	85	51	1967	284	123	1992	547	6
1943	86	55	1968	369	162			

(a) These data have been stored in a grouped file named ALUMNI.83g. Download this grouped file into your calculator. The lists are set up in ALUMNI.83g as CLASS, ROLL, and DONOR, respectively. Use the calculator to create a new variable which reports the percentage of each class' alumni who made a financial contribution to the college. Ignore the class year for the moment and use the calculator to analyze the distribution of donor percentages. Write a few sentences commenting on key features of this distribution. (Consider this your occasional reminder to *always* relate your comments to the context at hand.)

(b) Now use the calculator to produce a scatterplot of donor percentage vs. class. Comment on any patterns revealed in the scatterplot.

(c) Does one class stick out as deviating substantially from the *pattern* established by the majority? If so, identify it and suggest an explanation for its oddity.

Activity 7-12: Peanut Butter

The September 1990 issue of *Consumer Reports* rated 37 varieties of peanut butter. Each variety was given an overall sensory quality rating, based on taste tests by a trained sensory panel. Also listed was the cost (per three tablespoons, based on the average price paid by *CU* shoppers) and sodium content (per three tablespoons, in milligrams) of each product. Finally, each variety was classified as creamy (cr) or chunky (ch), natural (n) or regular (r), and salted (s) or unsalted (u). The results are given in the following table.

Brand	Cost	Sodium	Quality	Cr/ch	R/n	S/u
Jif	22	220	76	cr	r	s
Smucker's Natural	27	15	71	cr	n	u
Deaf Smith Arrowhead Mills	32	0	69	cr	n	u
Adams 100% Natural	26	0	60	cr	n	u
Adams	26	168	60	cr	n	s
Skippy	19	225	60	cr	r	s
Laura Scudder's All Natural	26	165	57	cr	n	s
Kroger	14	240	54	cr	r	s
Country Pure Brand	21	225	52	cr	n	s
NuMade	20	187	43	cr	r	s
Peter Pan	21	225	40	cr	r	s
Peter Pan	22	3	35	cr	r	u
A&P	12	225	34	cr	r	s
Hollywood Natural	32	15	34	cr	n	u
Food Club	17	225	33	cr	r	s
Pathmark	9	255	31	cr	r	s
Lady Lee	16	225	23	cr	r	s
Albertsons	17	225	23	cr	r	s
Shur Fine	16	225	11	cr	r	s
Smucker's Natural	27	15	89	ch	n	u
Jif	23	162	83	ch	r	s
Skippy	21	211	83	ch	r	s
Adams 100% Natural	26	0	69	ch	n	u
Deaf Smith Arrowhead Mills	32	0	69	ch	n	u

Brand	Cost	Sodium	Quality	Cr/ch	R/n	S/u
Country Pure Brand	21	195	67	ch	n	s
Laura Scudder's All Natural	24	165	63	ch	n	s
Smucker's Natural	26	188	57	ch	n	s
Food Club	17	195	54	ch	r	s
Kroger	14	255	49	ch	r	s
A&P	11	225	46	ch	r	s
Peter Pan	22	180	45	ch	r	s
NuMade	21	208	40	ch	r	s
Health Valley 100% Natural	34	3	40	ch	n	u
Lady Lee	16	225	34	ch	r	s
Albertsons	17	225	31	ch	r	s
Pathmark	9	210	29	ch	r	s
Shur Fine	16	195	26	ch	r	s

(a) Lists containing the above data have been stored in a grouped file called PBUTTER.83g. In part (c) you are asked to create a labeled scatterplot. In order for your calculator to produce a labeled scatterplot you need to use the program LBLSCAT.83p. Download both the grouped file and the program into your calculator. The lists you download are named COST, SODIU, QUAL, CRCH, RN, and SU, respectively. Since the calculator accepts only numeric values in its lists, the categorical lists contain only zeros and ones representing the different classifications within the category.

(b) Select any *pair* of measurement variables that you would like to examine. Use your calculator to produce a scatterplot of these variables, and write a paragraph of a few sentences commenting on the relationship between them. (Be sure to indicate which variables you choose).

(c) Now select *one* of the three categorical variables, and use the LBLSCAT program in your calculator to produce a labeled scatterplot of your two variables from (b) using this new variable for the labels. Comment on any effect of this categorical variable as well as on any other features of interest in the plot.

Activity 7-13: States' SAT Averages

The August 31, 1992, issue of *The Harrisburg Evening-News* lists the average SAT score for each of the fifty states and the percentage of high school seniors in the state who take the SAT test; these are reproduced below.

State	Avg SAT	% taking	State	Avg SAT	% taking
Alabama	996	8	Montana	988	24
Alaska	908	42	Nebraska	1018	11
Arizona	933	27	Nevada	922	27
Arkansas	990	6	New Hampshire	923	76
California	900	46	New Jersey	891	75
Colorado	960	29	New Mexico	996	12
Connecticut	900	79	New York	882	75
Delaware	895	66	North Carolina	855	57
Florida	884	50	North Dakota	1068	6
Georgia	842	65	Ohio	951	23
Hawaii	878	56	Oklahoma	1007	9
Idaho	963	17	Oregon	925	55
Illinois	1010	15	Pennsylvania	877	68
Indiana	868	58	Rhode Island	881	70
Iowa	1096	5	South Carolina	831	59
Kansas	1033	10	South Dakota	1040	6
Kentucky	988	11	Tennessee	1013	13
Louisiana	991	9	Texas	876	44
Maine	882	66	Utah	1041	5
Maryland	907	66	Vermont	897	69
Massachusetts	902	80	Virginia	893	63
Michigan	987	11	Washington	916	50
Minnesota	1053	10	West Virginia	924	17
Mississippi	1004	4	Wisconsin	1029	11
Missouri	1004	11	Wyoming	978	13

(a) Use the calculator to look at a scatterplot of these data. Write a paragraph describing the relationship between average SAT score and percentage of students taking the test. Include a reasonable explanation for the type of association that is apparent.

(b) Which state has the highest SAT average? Would you conclude that this state does the best job of educating its students? Which state has the lowest SAT average? Would you conclude that this state does the worst job of educating its students? Explain.

(c) How does *your* home state compare to the rest in terms of SAT average and percentage of students taking the test? (Identify the state also.)

Activity 7-14: Governors' Salaries *(cont.)*

Recall from Activity 6-11 the data concerning governors' salaries. The following table lists each state's average pay and governor's salary (as of 1993) and also indicates the state's region of the country (Northeast, Midwest, South, or West).

State	Region	Avg pay	Gov sal	State	Region	Avg pay	Gov sal
Alabama	S	20,468	81,151	Montana	W	17,895	55,502
Alaska	W	29,946	81,648	Nebraska	MW	18,577	65,000
Arizona	W	21,443	75,000	Nevada	W	22,358	90,000
Arkansas	S	18,204	60,000	New Hampshire	NE	22,609	86,235
California	W	26,180	120,000	New Jersey	NE	28,449	85,000
Colorado	W	22,908	70,000	New Mexico	W	19,347	90,000
Connecticut	NE	28,995	78,000	New York	NE	28,873	130,000
Delaware	S	24,423	95,000	North Carolina	S	20,220	123,300
Florida	S	21,032	101,764	North Dakota	MW	17,626	68,280
Georgia	S	22,114	91,092	Ohio	MW	22,843	115,752
Hawaii	W	23,167	94,780	Oklahoma	S	20,288	70,000
Idaho	W	18,991	75,000	Oregon	W	21,332	80,000
Illinois	MW	25,312	103,097	Pennsylvania	NE	23,457	105,000
Indiana	MW	21,699	77,200	Rhode Island	NE	22,388	69,900
Iowa	MW	19,224	76,700	South Carolina	S	19,669	101,959
Kansas	MW	20,238	76,476	South Dakota	MW	16,430	74,649
Kentucky	S	19,947	86,352	Tennessee	S	20,611	85,000
Louisiana	S	20,646	73,440	Texas	S	22,700	99,122
Maine	NE	20,154	70,000	Utah	W	20,074	70,000
Maryland	S	24,730	120,000	Vermont	NE	20,532	85,977
Massachusetts	NE	26,689	75,000	Virginia	S	22,750	110,000
Michigan	MW	25,376	112,025	Washington	W	22,646	121,000
Minnesota	MW	23,126	109,053	West Virginia	S	20,715	90,000
Mississippi	S	17,718	75,600	Wisconsin	MW	21,101	92,283
Missouri	MW	21,716	90,312	Wyoming	W	20,049	70,000

This labeled scatterplot displays governor's salary vs. average wage, using different letters for the regions (A=Northeast, B=Midwest, C=South, D=West).

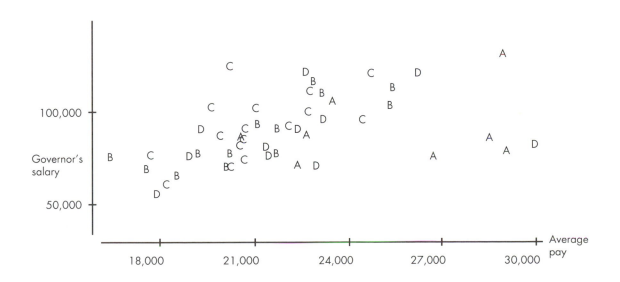

(a) Comment on the relationship between these two variables.
(b) List three pairs of states for which the state with the higher average pay has the lower governor salary.
(c) Name a state which appears to have a governor salary much higher than would be expected for a state with its average pay.
(d) Identify a cluster of four states which seem to have much lower governor salaries than would be expected for states with their average pay.

Activity 7-15: Teaching Evaluations

Investigate whether there seems to be an association between the number of students in a class and the students' average rating of the instructor on the course evaluation. The following table lists these variables for 25 courses taught by an instructor over a six-year period. The students' ratings of the instructor are on a scale of 1 to 9.

Course	Students	Avg rating	Course	Students	Avg rating	Course	Students	Avg rating
1	11	6.7	10	24	5.3	18	20	6.9
2	12	5.9	11	20	6.7	19	10	8.5
3	21	6.8	12	24	7.8	20	24	7.8
4	32	5.3	13	20	5.7	21	5	7.8
5	23	5.2	14	17	6.5	22	23	7.3
6	13	7.2	15	23	6.4	23	12	7.3
7	20	5	16	17	6.4	24	21	7
8	20	5.5	17	13	7.6	25	8	7.9
9	8	6.5						

Use the calculator to examine these data for evidence of an association between these two variables. If you detect an association, comment on its direction and strength.

Activity 7-16: Variables of Personal Interest *(cont.)*

Think of two *pairs* of measurement variables whose relationship you might be interested in studying. (Remember that a measurement variable is any characteristic of a person or object that can assume a range of numbers.) Be very specific in describing these variables; identify the *cases* as well as the variables.

WRAP-UP

This topic has introduced you to an area that occupies a prominent role in statistical practice: exploring relationships between variables. You have discovered the concept of *association* and learned how to construct and to interpret *scatterplots* and *labeled scatterplots* as visual displays of association.

Just as we moved from graphical displays to numerical summaries when describing distributions of data, in the next topic we will consider a numerical measure of the degree of association between two variables—the *correlation coefficient*.

Topic 8:

CORRELATION COEFFICIENT

OVERVIEW

You have seen how scatterplots provide useful visual information about the relationship between two variables. Just as you made use of numerical summaries of various aspects of the distribution of a single variable, it would also be handy to have a numerical measure of the association between two variables. This topic introduces you to just such a measure and asks you to investigate some of its properties. This measure is one of the most famous in statistics—the *correlation coefficient*.

OBJECTIVES

- To explore basic properties of the *correlation coefficient* as a numerical measure of the degree of association between two variables.
- To discover some of the limitations of the correlation coefficient as a summary of the relationship between two variables.
- To recognize the important distinction between association and *causation*.
- To become familiar with judging correlation values from scatterplots.
- To learn to use scatterplots and correlations to look for and to describe relationships between variables when analyzing genuine data.

PRELIMINARIES

1. Take a guess as to the number of people per television set in the United States in 1990. Do the same for China and for Haiti.

2. Do you expect that countries with few people per television tend to have longer life expectancies, shorter life expectancies, or do you suspect no relationship between televisions and life expectancy?

3. Taken from *An Altogether New Book of Top Ten Lists,* here are David Letterman's "Top Ten Good Things About Being a Really, Really Dumb Guy":

   A. Get to have own talk show with Canadian bandleader.
   B. G.E. executive dining room has great clam chowder.
   C. Already know the answer when people ask, "What are you—an idiot?"
   D. Can feel superior to really, really, really dumb guys.
   E. May get to be Vice-President of the United States.
   F. Pleasant sense of relief when Roadrunner gets away from Coyote.
   G. Fun bumper sticker: "I'd Rather Be Drooling."
   H. Never have to sit through long, boring Nobel Prize banquet.
   I. Stallone might play you in the movie.
   J. Seldom interrupted by annoying request to "Put that in layman's terms."

These have been presented in a randomly determined order. Rank these ten jokes from funniest (1) to least funny (10). Record your ranking of each joke next to its letter designation:

Joke letter	A	B	C	D	E	F	G	H	I	J
Your ranking										

IN-CLASS ACTIVITIES

The *correlation coefficient*, denoted by *r*, is a measure of the degree to which two variables are associated. The calculation of *r* is very tedious to do by hand, so you will begin by letting the calculator compute correlations while you explore their properties.

Activity 8-1: Properties of Correlation

(a) For this activity you will need to download lists containing the hypothetical exam scores and a program named CORR.83p. The lists containing the hypothetical exam scores are stored in a grouped file named HYPOCORR.83g. Download this grouped file and program into your calculator. The lists have been loaded as EXA1, EXA2, EXB1, EXB2, EXC1, EXC2, EXD1, EXD2, and so on.

In order for your calculator to compute the correlation coefficient you will need to turn the diagnostic display mode to "on". To do this, access the calculator's catalog (press ┃ 2nd ┃ ┃ 0 ┃) and scroll down until you find "DiagnosticOn." Press your ┃ENTER┃ button twice.

(b) Look back on the scatterplots of the six classes (A–F) of hypothetical exam scores that you examined in Topic 7. The table below indicates the direction and strength of the association in each case. Use CORR.83p to compute the correlation coefficient in each class and record its value in the table beside the appropriate letter designation. For example, with class A you want to find the correlation between the data in lists EXA1 and EXA2.

	Strong	Moderate	Least strong
Negative	C	D	F
Positive	E	A	B

(c) Based on these results, what do you suspect is the largest value that a correlation coefficient can assume? What do you suspect is the smallest value?

(d) Under what circumstances do you think the correlation assumes its largest or smallest value; i.e., what would have to be true of the observations in that case?

(e) How does the value of the correlation relate to the *direction* of the association?

(f) How does the value of the correlation relate to the *strength* of the association?

(g) Consider the scatterplot of another hypothetical data set below. Does there seem to be any relationship between the exam scores in class G? If so, describe the relationship.

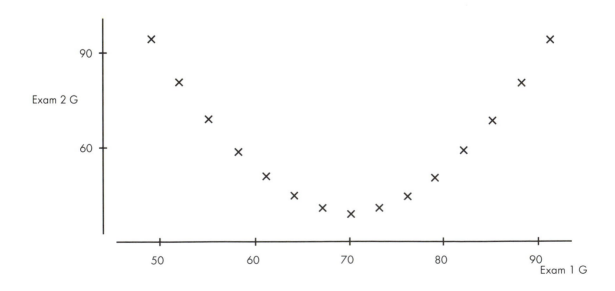

(h) Have the calculator compute the correlation coefficient for class G (lists EXG1 and EXG2); record it below. Does its value surprise you? Explain.

The example above illustrates that the correlation coefficient measures only *linear* (straight-line) relationships between two variables. More complicated types of relationships (such as curvilinear ones) can go undetected by *r*. Thus, there might be a relationship even if the correlation is close to zero. One should be aware of such possibilities and examine a scatterplot as well as the value of *r*.

Consider the scatterplots of two more hypothetical data sets:

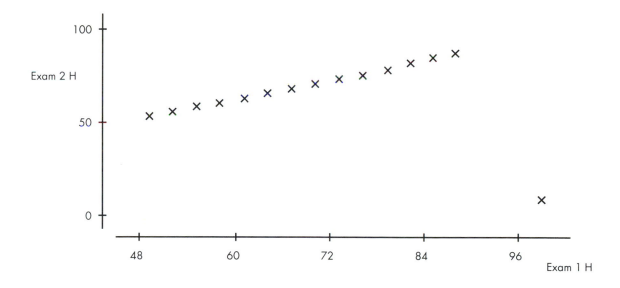

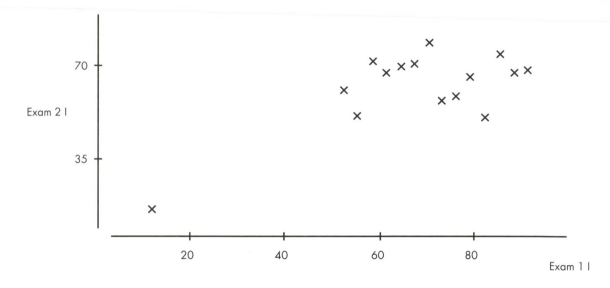

(i) In class H, do most of the observations seem to follow a linear pattern? Are there any exceptions?

(j) In class I, do most of the observations seem to be scattered haphazardly with no apparent pattern? Are there any exceptions?

(k) Have the calculator compute the correlation coefficient for each of these classes; record them below. Are you surprised at either of the values? Explain.

(l) Remove the outliers in classes H and I and have the calculator re-compute the correlation coefficients. Record them and comment on how they have changed.

(m) Based on your analyses of classes H and I, would you say that the correlation coefficient is a *resistant* measure of association? Explain.

Consider the scatterplot of one final hypothetical data set:

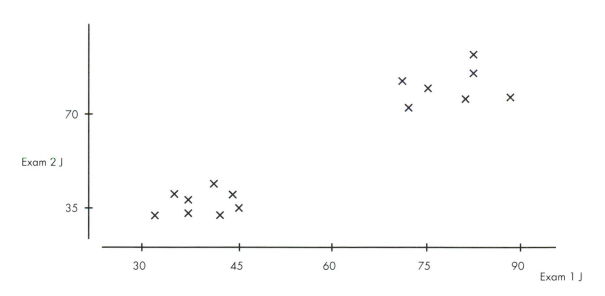

(n) Describe what the scatterplot reveals about the relationship between exam scores in class J.

(o) Have the calculator compute the correlation coefficient between exam scores in class J. Is its value higher than you expected?

Activity 8-2: Televisions and Life Expectancy

The following table provides information on life expectancies for a sample of 22 countries. It also lists the number of people per television set in each country.

Country	Life exp	Per TV	Country	Life exp	Per TV
Angola	44	200	Mexico	72	6.6
Australia	76.5	2	Morocco	64.5	21
Cambodia	49.5	177	Pakistan	56.5	73
Canada	76.5	1.7	Russia	69	3.2
China	70	8	South Africa	64	11
Egypt	60.5	15	Sri Lanka	71.5	28
France	78	2.6	Uganda	51	191
Haiti	53.5	234	United Kingdom	76	3
Iraq	67	18	United States	75.5	1.3
Japan	79	1.8	Vietnam	65	29
Madagascar	52.5	92	Yemen	50	38

(a) Which of the countries listed has the fewest people per television set? Which has the most? What are those numbers?

(b) These data have been stored in a grouped file named TVLIFE.83g. Download this grouped file into your calculator. The lists are labeled LFEXP and PERTV. Use the calculator to produce a scatterplot of life expectancy vs. people per television set. Does there appear to be an association between the two variables? Elaborate briefly.

(c) Have the calculator compute the value of the correlation coefficient between life expectancy and people per television.

(d) Since the association is so strongly negative, one might conclude that simply sending television sets to the countries with lower life expectancies would cause their inhabitants to live longer. Comment on this argument.

(e) If two variables have a correlation close to +1 or to −1, indicating a strong linear association between them, does it follow that there must be a cause-and-effect relationship between them?

This example illustrates the very important distinction between *association* and *causation*. Two variables may be strongly associated (as measured by the correlation coefficient) without a cause-and-effect relationship existing between them. Often the explanation is that both variables are related to a third variable not being measured; this variable is often called a *lurking* or *confounding variable.*

(f) In the case of life expectancy and television sets, suggest a confounding variable that is associated both with a country's life expectancy and with the prevalence of televisions in the country.

One can gain some insight into how the correlation coefficient *r* measures association by examining the formula for its calculation:

$$r = \frac{\sum_{i=1}^{n} \left(\frac{x_i - \bar{x}}{s_x}\right)\left(\frac{y_i - \bar{y}}{s_y}\right)}{n - 1},$$

where x_i denotes the ith observation of one variable, y_i the ith observation of the other variable, $\bar{x}$ and $\bar{y}$ the respective sample means, s_x and s_y the respective sample standard deviations, and n the sample size. This formula says to standardize each x and y value into its z-score, to multiply these z-scores together for each pair, to add those results, and to divide the sum by one less than the sample size.

Activity 8-3: Cars' Fuel Efficiency (*cont.*)

For the fuel efficiency data that you analyzed in Activity 7-1, the weights have a mean of 3208 and a standard deviation of 590. The mean of the miles per gallon ratings is 27.56, and their standard deviation is 6.70. The following table begins the process of calculating the correlation between weight and miles per gallon by calculating the z-scores for the weights and miles per gallons and then multiplying the results.

Model	Weight	Weight z-score	MPG	MPG z-score	Product
BMW 3-Series	3250	0.07	28	0.07	0.00
BMW 5-Series	3675	0.79	23	−0.68	−0.54
Cadillac Eldorado	3840	1.07	19	−1.28	−1.37
Cadillac Seville	3935	1.23	20	−1.13	−1.39
Ford Aspire	2140	−1.81	43	2.30	−4.17
Ford Crown Victoria	4010	1.36	22	−0.83	−1.13
Ford Escort	2565	−1.09	34	0.96	−1.05
Ford Mustang	3450	0.41	22	−0.83	−0.34
Ford Probe	2900	−0.52	28	0.07	−0.03
Ford Taurus	3345	0.23	25	−0.38	−0.09
Ford Taurus SHO	3545	0.57	24	−0.53	−0.30
Honda Accord	3050	−0.27	31	0.51	−0.14
Honda Civic	2540	−1.13	34	0.96	−1.09
Honda Civic del Sol	2410	−1.35	36	1.26	−1.70
Honda Prelude	2865		30	0.36	
Lincoln Mark VIII	3810	1.02	22		

(a) Calculate the z-score for the weight of a Honda Prelude and for the miles-per-gallon rating of a Lincoln Mark VIII. Show your calculations below and record the results in the table.

(b) Add the results of the "product" column and then divide the result by 15 (one less than the sample size of 16 cars) to determine the value of the correlation coefficient between weight and miles per gallon.

(c) What do you notice about the miles per gallon *z*-score of most of the cars with negative weight *z*-scores? Explain how this results from the strong negative association between weight and miles per gallon.

Activity 8-4: Guess the Correlation

This activity will give you practice at judging the value of a correlation co-efficient by examining the scatterplot.

(a) Download SCATSIM.83p into your calculator.

(b) Have the calculator generate some "pseudo-random" data and look at a scatterplot by running SCATSIM.83p. Based solely on the scatterplot, take a guess at the value of the correlation coefficient , recording your guess in the table below. (Make it a guessing contest with a partner!) The calculator computes the actual value of *r*; record it in the table also. Repeat this for a total of ten "pseudo-random" data sets.

repetition	1	2	3	4	5	6	7	8	9	10
guess										
actual										

(c) Before pressing ENTER to get your score (which is the correlation be-tween your guesses and actual values of *r*), make a guess as to what the value of the correlation coefficient between *your guesses* for *r* and the *ac-tual values* of *r* would be.

(d) Press ENTER and write down the value of the correlation coefficient between *your guesses* for *r* and the *actual values* of *r*. Are you surprised?

Homework Activities

Activity 8-5: Properties of Correlation (*cont.*)

Suppose that every student in a class scores ten points *higher* on the second exam than on the first exam.

(a) Produce (by hand) a rough sketch of what the scatterplot would look like.
(b) Enter some data which satisfy this condition into the calculator and compute the value of the correlation coefficient between the two exam scores.
(c) Repeat (a) and (b) supposing that every student scored 20 points *lower* on the second exam than on the first.
(d) Repeat (a) and (b) supposing that every student scored *twice* as many points on the second exam as on the first.
(e) Repeat (a) and (b) supposing that every student scored *one-half* as many points on the second exam as on the first.
(f) Based on your investigation of these questions, does the degree of the slope evident in a scatterplot relate to the correlation between the two variables?

Activity 8-6: States' SAT Averages *(cont.)*

Reconsider the data from Activity 7-13 concerning the percentage of high school seniors in a state who take the SAT test and the state's average SAT score.

(a) Use the calculator to compute the correlation coefficient between the percentage of high school seniors in a state who take the SAT and the average SAT score in that state.
(b) Would you conclude that a cause-and-effect relationship exists between these two variables? Explain.

Activity 8-7: Ice Cream, Drownings, and Fire Damage

(a) Many communities find a strong positive correlation between the amount of ice cream sold in a given month and the number of drownings that occur in that month. Does this mean that ice cream *causes* drowning? If not, can you think of an alternative explanation for the strong association? Write a few sentences addressing these questions.

(b) Explain why one would expect to find a positive correlation between the number of fire engines that respond to a fire and the amount of damage done in the fire. Does this mean that the damage would be less extensive if only fewer fire engines were dispatched? Explain.

Activity 8-8: Evaluation of Course Effectiveness

Suppose that a college professor has developed a new freshman course that he hopes will instill students with strong general learning skills. As a means of assessing the success of the course, he waits until a class of freshmen that have taken the course proceed to graduate from college. The professor then looks at two variables: the score on the final exam for his freshman course and cumulative college grade point average. Suppose that he finds a very strong positive association between the two variables (e.g., $r = 0.92$). Suppose further that he concludes that his course must have had a positive effect on students' learning skills, for those who did well in his course proceeded to do well in college; those who did poorly in his course went on to do poorly in college. Comment on the validity of the professor's conclusion.

Activity 8-9: Space Shuttle O-Ring Failures *(cont.)*

Reconsider the data from Activity 7-5 concerning space shuttle missions. Use the calculator to determine the value of the correlation between temperature and number of O-ring failures. Then exclude the flights in which no O-rings failed and recalculate the correlation. Explain why these correlations turn out to be so different.

Activity 8-10: Climatic Conditions *(cont.)*

The following table lists a number of climatic variables for a sample of 25 American cities. These variables measure long-term averages of:

- January high temperature (in degrees Fahrenheit)
- January low temperature
- July high temperature
- July low temperature
- annual precipitation (in inches)

- days of measurable precipitation per year
- annual snow accumulation
- percentage of sunshine

City	Jan. hi	Jan. lo	July hi	July lo	Precip.	Precip. days	Snow	Sun
Atlanta	50.4	31.5	88	69.5	50.77	115	2	61
Baltimore	40.2	23.4	87.2	66.8	40.76	113	21.3	57
Boston	35.7	21.6	81.8	65.1	41.51	126	40.7	58
Chicago	29	12.9	83.7	62.6	35.82	126	38.7	55
Cleveland	31.9	17.6	82.4	61.4	36.63	156	54.3	49
Dallas	54.1	32.7	96.5	74.1	33.7	78	2.9	64
Denver	43.2	16.1	88.2	58.6	15.4	89	59.8	70
Detroit	30.3	15.6	83.3	61.3	32.62	135	41.5	53
Houston	61	39.7	92.7	72.4	46.07	104	0.4	56
Kansas City	34.7	16.7	88.7	68.2	37.62	104	20	62
Los Angeles	65.7	47.8	75.3	62.8	12.01	35	0	73
Miami	75.2	59.2	89	76.2	55.91	129	0	73
Minneapolis	20.7	2.8	84	63.1	28.32	114	49.2	58
Nashville	45.9	26.5	89.5	68.9	47.3	119	10.6	56
New Orleans	60.8	41.8	90.6	73.1	61.88	114	0.2	60
New York	37.6	25.3	85.2	68.4	47.25	121	28.4	58
Philadelphia	37.9	22.8	82.6	67.2	41.41	117	21.3	56
Phoenix	65.9	41.2	105.9	81	7.66	36	0	86
Pittsburgh	33.7	18.5	82.6	61.6	36.85	154	42.8	46
St. Louis	37.7	20.8	89.3	70.4	37.51	111	19.9	57
Salt Lake City	36.4	19.3	92.2	63.7	16.18	90	57.8	66
San Diego	65.9	48.9	76.2	65.7	9.9	42	0	68
San Francisco	55.6	41.8	71.6	65.7	19.7	62	0	66
Seattle	45	35.2	75.2	55.2	37.19	156	12.3	46
Washington	42.3	26.8	88.5	71.4	38.63	112	17.1	56

(a) Use the calculator to compute the correlation coefficient between all pairs of these eight variables, recording your results in a table like the one below. (You need not record each value twice.)

	Jan. hi	Jan. lo	July hi	July lo	Precip.	Precip. day	Snow	Sun
Jan. hi	xxxx							
Jan. lo		xxxx						
July hi			xxxx					
July lo				xxxx				
Precip.					xxxx			
Prec. day						xxxx		
Snow							xxxx	
Sun								xxxx

(b) Which pair of variables has the *strongest* association? What is the correlation between them?

(c) Which pair of variables has the *weakest* association? What is the correlation between them?

(d) Suppose that you want to predict the annual snowfall for an American city and that you are allowed to look at that city's averages for these other variables. Which would be *most* useful to you? Which would be *least* useful?

(e) Suppose that you want to predict the average July high temperature for an American city and that you are allowed to look at that city's averages for these other variables. Which would be *most* useful to you? Which would be *least* useful?

(f) Use the calculator to explore the relationship between annual snowfall and annual precipitation more closely. Look at and comment on a scatterplot of these variables.

Activity 8-11: Students' Family Sizes *(cont.)*

Reconsider the data on students' family sizes collected in Topic 7 "Preliminaries." Use the calculator to compute the correlation coefficients for each of the three *pairs* of variables. Which pair has the highest (in absolute value) correlation?

Activity 8-12: Students' Travels *(cont.)*

Reconsider the data on students' travels collected in Topic 2. Use the calculator to look at a scatterplot of the number of states visited vs. number of countries visited and to calculate the correlation coefficient between these variables. Comment on what the correlation indicates about a relationship between the variables.

Activity 8-13: Students' Measurements *(cont.)*

Reconsider the data on students' measurements collected in Topic 3. Use the calculator to compute the correlation coefficients for each of the three *pairs* of variables. Which pair has the highest (in absolute value) correlation?

Activity 8-14: "Top Ten" Rankings

Recall your ranking of the "Top Ten" list from the "Preliminaries" section. David Letterman's ranking appears in the table below.

Joke letter	A	B	C	D	E	F	G	H	I	J
Your ranking										
Letterman's ranking	6	8	2	4	3	9	1	10	5	7

(a) Construct (by hand) a scatterplot of Letterman's ranking vs. your ranking of these jokes.

(b) Enter these rankings into the calculator and compute the value of the correlation between your ranking and Letterman's ranking. Record its value, and comment on how closely your ranking matches Dave's.

(c) Enter the rankings of your partner or another classmate into the calculator and compute the correlation between your rankings. Are your rankings more strongly correlated with Letterman's or with your classmate's?

Activity 8-15: *Star Trek* Episodes *(cont.)*

Refer back to the data concerning ranking *Star Trek* episodes in Activity 6-16.

(a) Based on the summary statistics and boxplots presented in Activity 6-16, would you expect an episode's ranking to be positively or negatively correlated with its chronological number?

(b) Use the calculator to examine a scatterplot of an episode's ranking vs. its chronological number. Does the scatterplot reveal any obvious association between the two?

(c) Have the calculator compute the correlation coefficient between an episode's ranking and its chronological number. Report the value of this correlation; does its sign agree with your answer to (a)?

Activity 8-16: Variables of Personal Interest *(cont.)*

Think of a situation in which you would expect two variables to be strongly correlated even though no cause-and-effect relationship exists between them. Describe these variables in a paragraph; also include an explanation for their strong association.

WRAP-UP

In this topic you have discovered the very important *correlation coefficient* as a measure of the linear relationship between two variables. You have derived some of the properties of this measure, such as the values it can assume, how its sign and value relate to the direction and strength of the association, and its lack of resistance to outliers. You have also practiced judging the direction and strength of a relationship from looking at a scatterplot. In addition, you have discovered the distinction between correlation and *causation* and learned that one needs to be very careful about inferring causal relationships between variables based solely on a strong correlation.

The next topic will expand your understanding of relationships between variables by introducing you to *least squares regression*, a formal mathematical model which is often useful for describing such relationships.

Topic 9:

LEAST SQUARES REGRESSION I

OVERVIEW

In previous topics you studied scatterplots as visual displays of the relationship between two variables and the correlation coefficient as a numerical measure of the linear association between them. With this topic you will begin to investigate *least squares regression* as a formal mathematical model often used to describe the relationship between variables.

OBJECTIVES

- To develop an awareness of *least squares regression* as a technique for modeling the relationship between two variables.
- To learn to use regression lines to make *predictions* and to recognize the limitations of those predictions.
- To understand some concepts associated with regression such as *fitted values*, *residuals*, and *proportion of variability explained*.
- To use the calculator to apply regression techniques with judgment and thoughtfulness to genuine data.

PRELIMINARIES

1. If one city is farther away than another, do you expect that it generally costs more to fly to the farther city?

2. Take a guess as to how much (on average) each additional mile adds to the cost of air fare.

3. Would you guess that distance explains about 50% of the variability in air fares, about 65% of this variability, or about 80% of this variability?

4. Take a guess concerning the highest yearly salary awarded to a rookie professional basketball player in 1991.

In-Class Activities

Activity 9-1: Air Fares (*cont.*)

Consider the data from Activity 7-7 concerning distances and air fares, a scatterplot of which appears below:

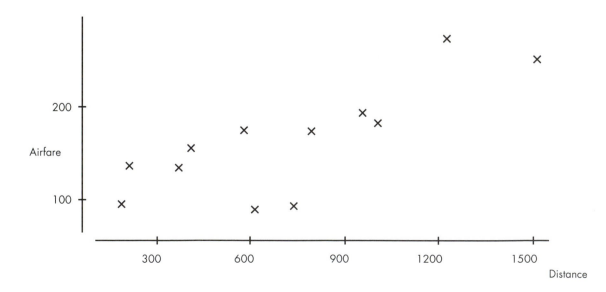

A natural goal is to try to use the distance of a destination to predict the air fare for flying there, and the simplest model for this prediction is to assume that a straight line summarizes the relationship between distance and air fare.

(a) Place a thread over the scatterplot above so that the thread forms a straight line which roughly summarizes the relationship between distance and air fare. Then draw this line on the scatterplot.

(b) Roughly what air fare does your line predict for a destination which is 300 miles away?

(c) Roughly what air fare does your line predict for a destination which is 1500 miles away?

The equation of a line can be represented as $y = a + bx$, where y denotes the variable being predicted (which is plotted on the vertical axis), x denotes the variable being used for the prediction (which is plotted on the horizontal axis), a is the value of the *y-intercept* of the line, and b is the value of the *slope* of the line. In this case x represents distance and y air-fare.

(d) Use your answers to (b) and (c) to find the *slope* of your line, remembering that since slope $= \dfrac{\text{rise}}{\text{run}} = \dfrac{\text{change in } y}{\text{change in } x}$, $b = \dfrac{y_2 - y_1}{x_2 - x_1}$.

(e) Use your answers to (d) and (b) to determine the *intercept* of your line, remembering that $a = y_1 - b \cdot x_1$.

(f) Put your answers to (d) and (e) together to produce the equation of your line. It is good form to replace the generic x and y symbols in the equation with the actual variable names, in this case *distance* and *air fare*, respectively.

Naturally, we would like to have a better way of choosing the line to approximate a relationship than simply drawing one that seems about right. Since there are infinitely many lines that one could draw, however, we need some criterion to select which line is the "best" at describing the relationship. The most commonly used criterion is *least squares*, which says to choose the line that minimizes the sum of squared vertical distances from the points to the line. We write the equation of the *least squares line* (also known as the *regression line*) as

$$y = a + bx,$$

where the *slope coefficient b* and the *intercept coefficient a* are determined from the sample data. The most convenient expression for calculating these coefficients relates them to the means and standard deviations of the two variables and the correlation coefficient between them:

$$b = r \frac{s_y}{s_x}, \qquad a = \bar{y} - b\bar{x},$$

where, you will recall, $\bar{x}$ and $\bar{y}$ represent the means of the variables, s_x and s_y their standard deviations, and r the correlation between them.

Activity 9-2: Air Fares (*cont.*)

(a) Enter the data from Activity 7-7 into your calculator, naming the lists DIST and AIRF, respectively.

(b) Use CORR.83p to compute the mean and standard deviation of distance and air fare, and the value of the correlation between the two. Record the results below:

	Mean	Std. dev.	Correlation
Airfare (y)			
Distance (x)			

(c) Use these statistics and the formulas given above to calculate the least squares coefficients *a* and *b*; record them below.

(d) Write the equation of the least squares line for predicting airfare from distance (using the variable names rather than the generic *x* and *y*). Then have the calculator determine this equation as a check on your calculations.

(e) To use your calculator to find the equation of the least squares line, press STAT ▷ 8 ENTER. You should have the follow ing displayed on your home screen.

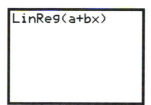

To find the least squares line for predicting air fare from distance, complete the screen above so that you have the following:

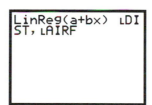

Press ENTER and write the equation below. Compare this equation with the equation you wrote down above. (Note that the calculator also gives the correlation coefficient).

One of the primary uses of regression is for *prediction*, for one can use the regression line (another name for the least squares line) to predict the value of the *y*-variable for a given value of the *x*-variable simply by plugging that value of *x* into the equation of the regression line. This is, of course, equivalent to finding the *y*-value of the point on the regression line corresponding to the *x*-value of interest.

(f) What airfare does the least squares line predict for a destination which is 300 miles away?

(g) What airfare does the least squares line predict for a destination which is 1500 miles away?

(h) Draw the least squares line on the scatterplot below by plotting the two points that you found in (f) and (g) and connecting them with a straight line.

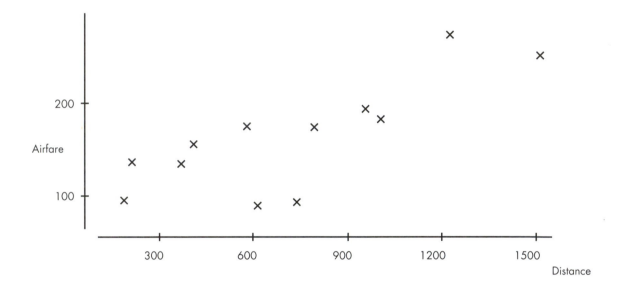

(i) Have your calculator create a scatterplot of airfare vs. distance and then graph the least squares line. To do this, enter the following onto your home screen (you can find Y₁ by pressing VARS ▷ 1):

Now press ENTER and GRAPH . The linear regression equation is located at the "Y₁=" prompt in the "Y=" editor.

(j) Just from looking at the regression line that you have drawn on the scatterplot, guess what value the regression line would predict for the airfare to a destination 900 miles away.

(k) Use the equation of the regression line to predict the airfare to a destination 900 miles away, and compare this prediction to your guess in (j).

(l) What airfare would the regression line predict for a flight to San Francisco, which is 2842 miles from Baltimore? Would you take this prediction as seriously as the one for 900 miles? Explain.

The actual airfare to San Francisco at that time was $198. That the regression line's prediction is not very reasonable illustrates the danger of *extrapolation*; i.e., of trying to predict y for values of x beyond those contained in the data. Since we have no reason to believe that the relationship between distance and airfare remains roughly linear beyond the range of values contained in our data set, such extrapolation is not advisable.

(m) Use the equation of the regression line to predict the airfare if the distance is 900 miles. Record the prediction in the table below, and repeat for distances of 901, 902, and 903 miles.

Distance	900	901	902	903
Predicted airfare				

(n) Do you notice a pattern in these predictions? By how many dollars is each prediction higher than the preceding one? Does this number look familiar (from your earlier calculations)? Explain.

This exercise demonstrates that one can *interpret* the slope coefficient of the least squares line as the predicted change in the *y*-variable (airfare, in this case) for a one-unit change in the *x*-variable (distance).

(o) By how much does the regression line predict airfare to rise for each additional 100 miles that a destination is farther away?

A common theme in statistical modeling is to think of each data point as being comprised of two parts: the part that is explained by the model (often called the *fit*) and the "leftover" part (often called the *residual*) that is either the result of chance variation or of variables not measured. In the context of least squares regression, the *fitted value* for an observation is simply the *y*-value that the regression line would predict for the *x*-value of that observation. The *residual* is the difference between the actual *y*-value and the fitted value (residual = actual − fitted), so the residual measures the vertical distance from the observed *y*-value to the regression line.

(p) If you look back at the original listing of distances and airfares, you find that Atlanta is 576 miles from Baltimore. What airfare would the regression line have predicted for Atlanta? (This is the *fitted value* for Atlanta.)

(q) The actual airfare to Atlanta at that time was $178. Determine the *residual* value for Atlanta by subtracting the predicted fare from the actual one.

(r) Record your answers to (p) and (q) in the following table (at the top of the next page). Then calculate Boston's residual and Chicago's fitted value *without* using the equation of the regression line, showing your calculations.

Destination	Distance	Airfare	Fitted	Residual
Atlanta	576	178		
Boston	370	138	126.70	
Chicago	612	94		−61.10
Dallas/Fort Worth	1216	278	226.00	52.00
Detroit	409	158	131.27	26.73
Denver	1502	258	259.57	−1.56
Miami	946	198	194.30	3.70
New Orleans	998	188	200.41	−12.41
New York	189	98	105.45	−7.45
Orlando	787	179	175.64	3.36
Pittsburgh	210	138	107.92	30.08
St. Louis	737	98	169.77	−71.77

(s) Which city has the largest (in absolute value) residual? What were its distance and airfare? By how much did the regression line err in predicting its airfare; was it an underestimate or an overestimate? Circle this observation on the scatterplot containing the regression line above.

(t) For observations with *positive* residual values, was their actual air fare greater or less than the predicted airfare?

(u) For observations with *negative* residual values, do their points on the scatterplot fall above or below the regression line?

(v) Use the calculator to compute summary statistics concerning the residuals. Record the mean and standard deviation of the residuals below.

Recall the idea of thinking of the data as consisting of a part explained by the statistical model and a "leftover" (residual) part due to chance variation or unmeasured variables. One can obtain a numerical measure of how much of the variability in the data is explained by the model and how much is "left over" by comparing the residual standard deviation with the standard deviation of the dependent (y-) variable.

(w) Recall that you found the standard deviation of the airfares in (a) and the standard deviation of the regression residuals in (s). Divide the standard deviation of the regression residuals by the standard deviation of the airfares. Then square this value and record the result below.

(x) Recall that you also found the correlation between distance and airfare in (a). Square this value and record the result below.

(y) What do you notice about the *sum* of your answers in (t) and (u)?

It turns out that the square of the correlation coefficient (written r^2) measures the *proportion of variability* in the y-variable that is explained by the regression model with the x-variable. This percentage measures how closely the points fall to the least squares line and thus also provides an indication of how confident one can be of predictions made with the line.

(z) What proportion of the variability in airfares is explained by the regression line with distance? (You have already done this calculation above.)

Activity 9-3: Students' Measurements (*cont.*)

Refer back to the data collected in Topic 3 on students' heights and foot length.

(a) Use the calculator to produce a scatterplot of height vs. foot length. Based on the scatterplot, does it look like a line would summarize well the relationship between height and foot length?

(b) Have the calculator determine the equation of the least squares line for predicting height from foot length. Record the equation below and have the calculator draw the line on the scatterplot. Does the line seem to summarize the relationship well?

(c) Interpret the value of the slope coefficient. In other words, explain what the value of the slope signifies in terms of predicting height from foot length.

(d) What proportion of the variability in heights is explained by the least squares line involving foot lengths?

(e) Use the equation of the least squares line to predict the height of a student whose foot length is 25 cm. Show your calculation below.

(f) Does it make sense to interpret the intercept coefficient in this case? In other words, is it sensible to predict the height of a person whose foot length is 0 cm?

HOMEWORK ACTIVITIES

Activity 9-4: Students' Measurements (*cont.*)

Consider again the data on students' heights and foot sizes, but this time analyze men's and women's data separately.

(a) Use the calculator to determine the regression equation for predicting a *male* student's height from his foot length. Record the equation.

(b) Use the calculator to determine the regression equation for predicting a *female* student's height from her foot length. Record the equation.

(c) Compare the slopes of these two equations in (a) and (b), commenting on any differences between them.

(d) What height would the regression equation in (a) predict for a man with a foot length of 25 cm?

(e) What height would the regression equation in (b) predict for a woman whose foot length is 25 cm?

(f) What height would the regression equation in Activity 9-3 predict for a student of unknown gender with a foot length of 25 cm?

(g) Comment on how much the predictions in (d), (e), and (f) differ.

Activity 9-5: Cars' Fuel Efficiency (*cont.*)

Refer back to the data in Activity 7-1 concerning the relationship between cars' weight and fuel efficiency. The means and standard deviations of these variables and the correlation between them are reported below:

	Mean	Std. dev.	Correlation
Weight	3208	590	−0.959
MPG	27.56	6.70	

(a) Use this information to determine (by hand) the coefficients of the least squares line for predicting a car's miles per gallon rating from its weight.

(b) By how many miles per gallon does the least squares line predict a car's fuel efficiency will drop for each additional 100 pounds of weight? (Use the slope coefficient to answer this question.)

(c) What proportion of the variability in cars' miles per gallon ratings is explained by the least squares line with weight?

Activity 9-6: Governors' Salaries *(cont.)*

Reconsider the data from Activity 7-14 concerning governors' salaries and average pay in the states.

(a) Use the calculator to determine the regression equation for predicting a state's governor's salary from its average pay. Also have the calculator compute residuals. Record the equation of the regression line.

(b) What proportion of the variability in governors' salaries is explained by this regression line with average pay?

(c) Use the calculator to compute residuals and fitted values. The calculator automatically calculates and enters the residuals into a list named RESID every time you use the calculator to find the linear regression equation. (Note: You can calculate the fitted values by using the equation: residual = actual − fitted).

(d) Which state has the largest positive residual? Explain what this signifies about the state.

(e) Which state has the largest (in absolute value) negative residual? Explain what this signifies about the state.

(f) Which state has the largest fitted value? Explain how you could determine this from the raw data without actually calculating any fitted values.

Activity 9-7: Basketball Rookie Salaries

The following table pertains to basketball players selected in the first round of the 1991 National Basketball Association draft. It lists the draft number (the order in which the player was selected) of each player and the annual salary of the contract that the player signed. The two missing entries are for players who signed with European teams.

Pick no.	Salary	Pick no.	Salary	Pick no.	Salary
1	$3,333,333	10	$1,010,652	19	$828,750
2	$2,900,000	11	$997,120	20	$740,000
3	$2,867,100	12	$1,370,000	21	$775,000
4	$2,750,000	13	$817,000	22	$180,000
5	$2,458,333	14	$675,000	23	$550,000
6	$1,736,250	15	*	24	$610,000
7	$1,590,000	16	$1,120,000	25	*
8	$1,500,000	17	$1,120,000	26	$180,000
9	$1,400,000	18	$875,000	27	$605,000

(a) Use the calculator to look at a scatterplot of the data and to compute the regression line for predicting salary from draft number. Record this regression equation.

(b) What proportion of the variability in salary is explained by the regression model with draft number?

(c) Calculate (by hand) the fitted value and residual for the player who was draft number 12.

(d) What yearly salary would the regression line predict for the player drafted at number 15? How about for number 25?

(e) By how much does the regression line predict the salary to drop for each additional draft number? In other words, how much does a player stand to lose for each additional draft position which passes him by?

Activity 9-8: Fast Food Sandwiches *(cont.)*

Reconsider the data from Activity 7-4 about fast food sandwiches. The mean serving size is 7.557 ounces; the standard deviation of serving sizes is 2.008 ounces. The mean calories per sandwich is 446.9 with a standard deviation of 143.0. The correlation between serving size and calories is 0.849.

(a) Use this information to determine the least squares line for predicting calories from serving size. Record the equation of this line.

(b) Reproduce the labeled scatterplot from Activity 7-4 and sketch the least squares line on it. How well does the line appear to summarize the relationship between calories and serving size?

(c) What proportion of the variability in calories is explained by the least squares line with serving size?

(d) Do the roast beef sandwiches tend to fall above the line, below the line, or about half and half? What about the chicken sandwiches? How about turkey? Comment on what your findings reveal.

Activity 9-9: Electricity Bills

The following table (at the top of the next page) lists the average temperature for a month and the amount of the electricity bill for that month. Notice that data are missing for three of the months.

Month	Temp	Bill	Month	Temp	Bill
Apr-91	51	$41.69	Jun-92	66	$40.89
May-91	61	$42.64	Jul-92	72	$40.89
Jun-91	74	$36.62	Aug-92	72	$41.39
Jul-91	77	$40.70	Sep-92	70	$38.31
Aug-91	78	$38.49	Oct-92	*	*
Sep-91	74	$37.88	Nov-92	45	$43.82
Oct-91	59	$35.94	Dec-92	39	$44.41
Nov-91	48	$39.34	Jan-93	35	$46.24
Dec-91	44	$49.66	Feb-93	*	*
Jan-92	34	$55.49	Mar-93	30	$50.80
Feb-92	32	$47.81	Apr-93	49	$47.64
Mar-92	41	$44.43	May-93	*	*
Apr-92	43	$48.87	Jun-93	68	$38.70
May-92	57	$39.48	Jul-93	78	$47.47

(a) Before you examine the relationship between average temperature and electric bill, examine the distribution of electric bill charges themselves. Create (by hand) a dotplot of the electric bill charges, and use the calculator to compute relevant summary statistics. Then write a few sentences describing the distribution of electric bill charges.

(b) Use the calculator to produce a scatterplot of electric bill vs. average temperature. Does the scatterplot reveal a positive association between these variables, a negative association, or not much association at all? If there is an association, how strong is it?

(c) Use the calculator to determine the equation of the least squares (regression) line for predicting the electric bill from the average temperature. Record the equation of the line.

(d) Use this equation to determine (by hand) the fitted value and residual for March of 1992.

(e) Use the calculator to compute the residuals and fitted values for each month. The calculator automatically calculates and enters the residuals into a list named RESID every time you use the calculator to find the linear regression equation. (Note: You can calculate the fitted values by using the equation: residual = actual − fitted). Which month(s) have unusually large residual values? Were their electric bills higher or lower than expected for their average temperature?

(f) Create (by hand) a dotplot of the distribution of residuals and comment on its key features.

Activity 9-10: Turnpike Tolls

If one enters the Pennsylvania Turnpike at the Ohio border and travels east to New Jersey, the mileages and tolls for the turnpike exits are as displayed in the scatterplot below. The regression line for predicting the toll from the mileage has been drawn on the scatterplot; its equation is: toll = $-0.123 + 0.0402$ mileage. The correlation between toll and mileage is 0.999.

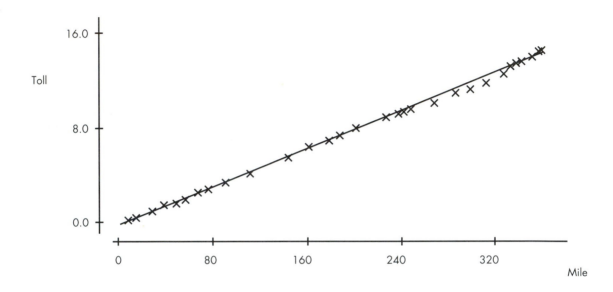

(a) What proportion of the variability in turnpike tolls is explained by the regression line with mileage?

(b) Use the regression equation to predict the toll for a person who needs to drive 150 miles on the turnpike.

(c) By how much does the regression equation predict the toll to rise for each additional mile that you drive on the turnpike?

(d) About how many miles do you have to drive in order for the toll to increase by one dollar?

Activity 9-11: Broadway Shows *(cont.)*

Refer to the data in Activity 2-7 concerning Broadway shows. It turns out that the least squares line for predicting receipts from percentage capacity is: receipts = $-265,892 + 8118$ (% capacity).

(a) What would the least squares line predict for the receipts of a show that filled 80% of its theater's capacity?
(b) Calculate the fitted value and residual for *Cats*.
(c) Which show has the largest positive residual?
(d) Which show has the smallest fitted value?

Activity 9-12: Climatic Conditions *(cont.)*

Reconsider the climatic data presented in Activity 8-10.

(a) Choose any *pair* of variables (preferably a pair which is strongly correlated) and use the calculator to determine the least squares line for predicting one variable from the other. Record the equation of this line, being sure to identify which variables you are considering.
(b) Select one particular value of the independent variable and use the least squares line to predicting the value of the dependent variable.
(c) Which city has the largest (in absolute value) residual from your least squares line? What is the value of this residual? Interpret what this residual says about the city.
(d) What proportion of the variability in the dependent variable is explained by the least squares line?

WRAP-UP

This topic has led you to study a formal mathematical model for describing the relationship between two variables. In studying *least squares regression*, you have encountered a variety of related terms and concepts. These ideas include the use of regression in *prediction*, the danger of *extrapolation*, the interpretation of the *slope coefficient*, the concepts of *fitted values* and *residuals*, and the interpretation of r^2 as the *proportion of variability explained* by the regression line. Understanding all of these ideas is important to applying regression techniques thoughtfully and appropriately.

In Topic 9 you will continue your study of least squares regression.

LEAST SQUARES REGRESSION II

OVERVIEW

This topic extends your study of least squares regression. You will examine the impact that a single observation can have on a regression analysis, learn how to use *residual plots* to indicate when the linear relationship is not appropriate, and discover *transformation* of variables as a way to use regression even when the relationship between the variables is not linear.

OBJECTIVES

- To understand the distinction and importance of *outliers* and *influential observations* in the context of regression analysis.
- To learn to use *residual plots* to indicate when the linear relationship is not a satisfactory model for describing the relationship between two variables.
- To discover how to *transform* variables to create a linear relationship between variables.

PRELIMINARIES

1. Identify an animal that has an especially long lifetime.

2. If one animal tends to live longer than another, would you expect it to have a longer gestation period than the other?

3. Take a guess concerning the average lifetime and gestation period of cats, horses, and rabbits.

IN-CLASS ACTIVITIES

Activity 10-1: Gestation and Longevity

The following table lists the average longevity (in years) and gestation period (in days) for a sample of animals, as reported in *The 1993 World Almanac and Book of Facts.*

Animal	Gestation	Longevity	Animal	Gestation	Longevity
Ass	365	12	Guinea pig	68	4
Baboon	187	20	Hippopotamus	238	25
Bear, black	219	18	Horse	330	20
Bear, grizzly	225	25	Kangaroo	42	7
Bear, polar	240	20	Leopard	98	12
Beaver	122	5	Lion	100	15
Buffalo	278	15	Monkey	164	15
Camel	406	12	Moose	240	12
Cat	63	12	Mouse	21	3
Chimpanzee	231	20	Opossum	15	1
Chipmunk	31	6	Pig	112	10
Cow	284	15	Puma	90	12
Deer	201	8	Rabbit	31	5
Dog	61	12	Rhinoceros	450	15
Elephant	645	40	Sea lion	350	12
Elk	250	15	Sheep	154	12
Fox	52	7	Squirrel	44	10
Giraffe	425	10	Tiger	105	16
Goat	151	8	Wolf	63	5
Gorilla	257	20	Zebra	365	15

(a) Load the grouped file ANIMALS.83g into your calculator. The list names are GEST and LONG, respectively. Use the calculator to determine the regression line for predicting an animal's gestation period from its longevity. Use the calculator to compute the residuals and fitted values. (The calculator automatically calculates and enters the residuals into a list named RESID every time you use the calculator to find the linear regression equation. Note that you can calculate the fitted values by using the equation: residual = actual − fitted). Record the equation of the line below and sketch it on the following scatterplot.

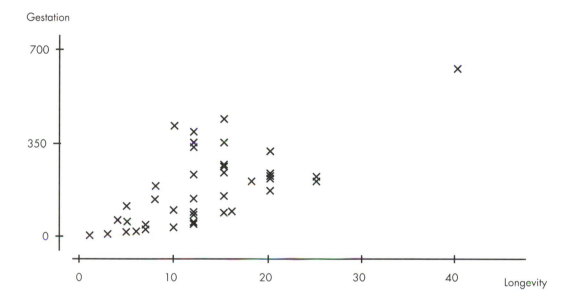

(b) Interpret the slope coefficient of the regression line. In other words, explain precisely what its value signifies about the relationship between longevity and gestation period.

(c) What proportion of the variability in animals' gestation periods is explained by the regression line?

(d) Use the calculator to create a scatterplot of the animals' residual values vs. their longevities (the residuals are in the list named **RESID**). Is there any relationship between residuals and longevities? If so, explain in a sentence or two what that signifies about the accuracy of predictions for animals with long vs. short lifetimes.

(e) Which of the animals is clearly an outlier both in longevity and in gestation period? Circle this animal on the scatterplot and calculate its residual value. Does it seem to have the largest residual (in absolute value) of any animal?

In the context of regression lines, *outliers* are observations with large (in absolute value) residuals. In other words, outliers fall far from the regression line, not following the pattern of the relationship apparent in the others. While the elephant is an outlier in both the longevity and gestation variables, it is not an outlier in the regression context.

(f) Which animal does have the largest (in absolute value) residual? Is its gestation period longer or shorter than expected for an animal with its longevity?

(g) Eliminate for the moment the giraffe's information from the analysis. Use the calculator to determine the equation of the regression line for predicting gestation period from longevity in this case. Record the equation and also the value of r^2 below.

(h) The following scatterplot displays the original data and both regression lines. Is this new regression line substantially different from the actual one?

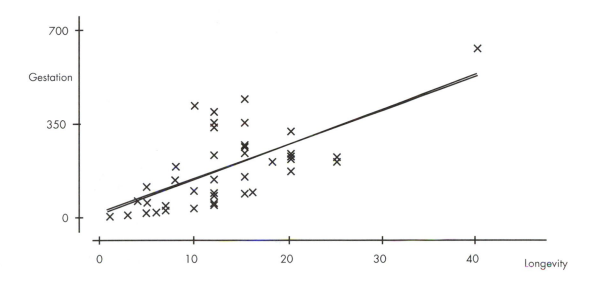

(i) Return the giraffe's values to the analysis but eliminate for now the elephant's. Use the calculator to determine the equation of the regression line in this case. Record this equation along with the value of r^2 below.

(j) The following scatterplot (on the next page) displays the original data with the actual regression line and the line found when the elephant had been removed. In which case (giraffe or elephant) did the removal of one animal affect the regression line more?

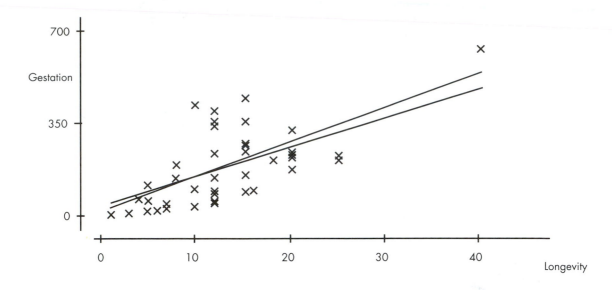

In the context of least squares regression, an *influential observation* is an observation which has a great influence on the regression line by virtue of its extreme *x*-value (longevity, in this case). Removing the elephant from the analysis has a considerable effect on the least squares line, whereas the removal of the giraffe does not. The elephant is an influential observation due to its exceptionally long lifetime.

(k) To appreciate even further the influence of the elephant, change its gestation period to 45 days instead of 645 days. Use the calculator to recompute the equation of the regression line and of r^2, recording each below. The following scatterplot reveals the new data along with both the original and revised regression lines.

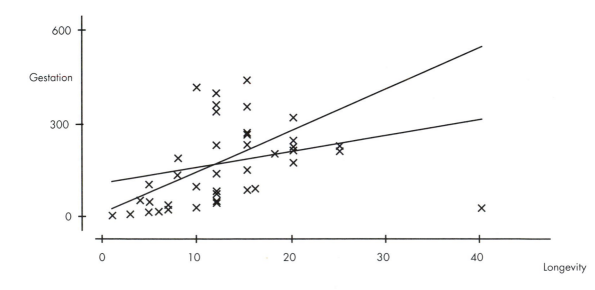

Activity 10-2: Planetary Measurements (*cont.*)

Reconsider the data from Activity 4-6 concerning planetary measurements. A scatterplot of the relationship between the planet's distance from the sun and its position among the planets appears below. (Mercury has position 1, Venus position 2, and so on through Pluto at position 9.)

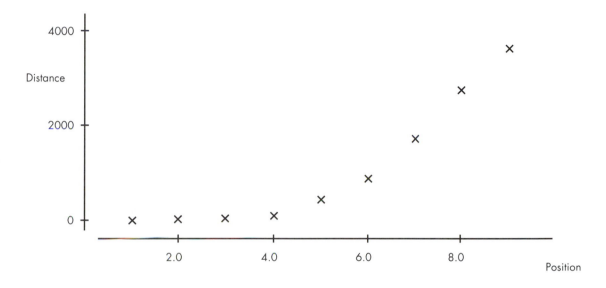

(a) Describe the relationship between distance and position in a sentence or two.

(b) Have the calculator compute the value of the correlation coefficient between a planet's distance from the sun and its position. Report this value and comment on its magnitude. Does the correlation's value seem to indicate a strong linear relationship between distance and position?

(c) Use the calculator to determine the least squares line for predicting a planet's distance from the sun based on its position. Record the equation of this line below and sketch it on the scatterplot above.

(d) By analyzing your scatterplot and the regression line, would you say that the line "fits" the data well? Explain.

(e) Use the calculator to compute the residuals from the regression line and then create (by hand) a scatterplot of residuals vs. position. Does this scatterplot reveal any pattern? Explain.

In this planetary example, the original scatterplot reveals very clearly that a straight line is not an accurate model for describing the relationship between the variables. In less obvious cases, a residual plot provides a useful check of the appropriateness of the linear model. When a straight line is a reasonable model, the residual plot should reveal a seemingly random scattering of points. When a nonlinear model would fit the data better, the residual plot reveals a pattern of some kind.

When a straight line is not the best mathematical description of a relationship, one can *transform* one or both variables to make the association more linear.

Activity 10-3: Televisions and Life Expectancy (*cont.*)

Reconsider the data from Activity 8-2 dealing with televisions and life expectancies.

(a) Use the calculator to look at a scatterplot of life expectancy vs. people per television and to calculate the correlation between these two. Describe the relationship between the variables; in particular, does it appear to be linear?

(b) Use the calculator to create a new variable: logarithm of the number of people per television. (You may use any base to take the logarithm, but base 10 is a natural choice.) To do this, type the following onto your home screen:

```
log(PERTV)→LPRTV
```

Now press ENTER.

(c) Have the calculator produce a scatterplot of life expectancy vs. the log of the number of people per television. Also calculate the correlation between these variables. Does the relationship appear to be stronger and more linear than before?

(d) Use the calculator to compute the equation of the least squares line for predicting a country's life expectancy from the log of its number of people per television. Record the equation below.

(e) What proportion of the variability in countries' life expectancies is explained by the regression equation with the log of the people per television?

(f) What life expectancy would this regression line predict for a country with 10 people per television? Take the log of the number of people per television and then plug that log value into the regression equation.

(g) Examine a scatterplot of the residuals of this regression equation vs. the log of the people per television. Does the scatterplot reveal any clear pattern?

(h) Is the linear regression model a better fit with the original data (i.e., from Activity 8-2) or with the transformed data?

HOMEWORK ACTIVITIES

Activity 10-4: Planetary Measurements (*cont.*)

Return to the analysis of Activity 10-2 in which you discovered that a least squares line is not a good fit for the relationship between a planet's distance from the sun and its position number.

(a) Use the calculator to create two new variables: square root of distance and logarithm of distance. Look at scatterplots of each of these variables vs. position. Which transformation (square root or logarithm) seems to produce a more linear relationship?

(b) For whichever transformation you select in (a) as more appropriate, use the calculator to determine the regression equation for predicting that transformation of distance from position. Record this equation, indicating which transformation you work with.

(c) Report the value of r^2 for this regression equation.

(d) Look at a residual plot and comment on whether the residuals seem to be scattered randomly or to follow a pattern.

Activity 10-5: Summer Blockbuster Movies *(cont.)*

Refer back to the data in Activity 3-10 concerning box office revenues for movies.

(a) Have the calculator produce a scatterplot of second week's box office revenue vs. first week's. Also calculate the correlation between these two. Describe any association that you find.

(b) Use the calculator to determine the regression equation for predicting a movie's second week box office revenue from its first week's box office revenue; report this equation. Also have the calculator compute residuals and fitted values.

(c) Which movie has the largest (in absolute value) residual? What is the value of this residual? What, specifically, is unusual about this movie?

(d) Examine a dotplot of the residuals. Report the mean and median of the residuals. How many and what proportion of the residuals are negative? Does the regression line seem to be overestimating or underestimating the second week box office revenue for most movies?

(e) Which movie is most likely to be an influential observation? Remove this movie from the analysis and have the calculator recompute the regression equation, residuals, and fitted values. Report the regression equation. Does it seem to have changed much?

Activity 10-6: College Enrollments

The following data (in the table on the next page) are the enrollments and faculty sizes of a sample of 34 American colleges and universities, taken from *The 1991 World Almanac and Book of Facts.*

College	Enrollment	Faculty	College	Enrollment	Faculty
1	12385	600	18	1671	114
2	1223	110	19	3778	234
3	2920	241	20	1250	71
4	1697	147	21	20110	733
5	890	113	22	16239	814
6	1131	87	23	766	63
7	1257	89	24	1875	115
8	2595	179	25	4170	390
9	52895	3796	26	1317	162
10	16500	950	27	3023	144
11	1080	45	28	15958	2362
12	1501	67	29	1220	44
13	1477	103	30	794	193
14	1369	114	31	2315	162
15	1385	49	32	1316	101
16	1966	160	33	800	63
17	2173	128	34	2381	144

(a) Use the calculator to produce a scatterplot of these data and to compute the correlation coefficient between enrollment and faculty size. Write a few sentences commenting on whether there is an association between the two variables and, if so, on the direction and strength of the association.

(b) Suppose that you want to use enrollment information to predict the number of faculty members that a college has. Use the calculator to determine the least squares line, recording its equation (expressed in terms of the relevant variables).

(c) *Interpret* the slope coefficient of the least squares line.

(d) Comment on whether there seem to be any outliers or influential observations in the sample. Identify such colleges by number and explain your answers.

Activity 10-7: Gestation and Longevity *(cont.)*

Recall the data that you analyzed above concerning the relationship between an animal's longevity and its gestation period.

(a) Use the regression equation that you found above to predict the gestation period of a human being. (You will need to estimate for yourself or discover in a reference source the average longevity of a human.) Show the details of your calculation.

(b) Do you accept this prediction as a reasonable one? If not, explain why the regression equation does not produce a reasonable prediction in this instance.

Activity 10-8: Turnpike Tolls *(cont.)*

Refer back to the regression analysis of Pennsylvania Turnpike tolls presented in Activity 9-10. A scatterplot of the regression residuals vs. mileage is given below:

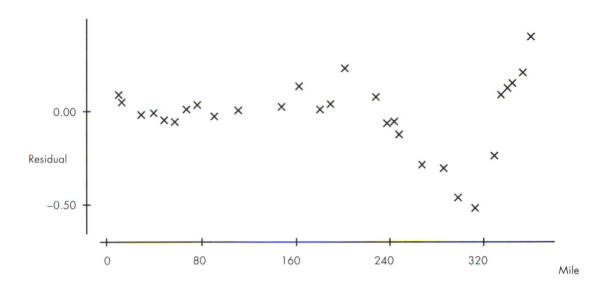

(a) Do the residuals seem to be randomly scattered or do some definite patterns emerge in this scatterplot?

(b) Despite the very high correlation of 0.999, does this residual plot give you cause to suspect that the relationship between toll and mileage is not completely linear? Explain. (You may also want to refer back to the scatterplot of toll vs. mileage in Activity 9-10.)

WRAP-UP

This topic has extended your study of least squares regression. You have examined *outliers* and *influential observations*, noting the effects that they can have on regression analysis. You have also learned to use *residual plots* to judge whether a nonlinear model might better describe the relationship between two variables, and you have discovered how to *transform* variables when such a nonlinear model is called for.

In the next topic you will continue your study of relationships between variables but with very different methods of analysis. You will turn your attention from measurement variables to categorical variables.

Topic 11:

RELATIONSHIPS WITH CATEGORICAL VARIABLES

OVERVIEW

You have been studying how to explore relationships between two variables, but the variables that you have analyzed have all been *measurement* variables. This topic asks you to study some basic techniques for exploring relationships among *categorical* variables. (Remember that a categorical variable is one which records simply that category into which a person or thing falls on the characteristic in question.) These techniques involve the analysis of two-way tables of counts. You will use no more complicated mathematical operations than addition and calculation of proportions, but you will acquire some very powerful analytical tools.

OBJECTIVES

- To learn to produce a *two-way table* as a summary of the information contained in a pair of categorical variables.
- To develop skills of interpreting information presented in two-way tables of counts.
- To become familiar with the concepts of *marginal* and *conditional distributions* of categorical variables.
- To discover *segmented bar graphs* as visual representations of the information contained in such tables.

- To acquire the ability to understand, recognize, and explain the phenomenon of *Simpson's paradox* as it relates to interpreting and drawing conclusions from two-way tables.
- To gain experience in applying techniques of analyzing two-way tables to genuine data.

PRELIMINARIES

1. Do you think that a student's political leaning has any bearing on whether he/she wants to see the penny retained or abolished?

2. Do you think that Americans tend to grow more liberal or more conservative in their political ideology as they grow older, or do you suspect that there is no relationship between age and political ideology?

3. Record for each student in the class whether he/she has read the novel *Jurassic Park* by Michael Crichton and whether he/she has seen the movie *Jurassic Park* directed by Steven Spielberg.

Stu-dent	Read book?	Seen movie?	Stu-dent	Read book?	Seen movie?	Stu-dent	Read book?	Seen movie?
1			9			17		
2			10			18		
3			11			19		
4			12			20		
5			13			21		
6			14			22		
7			15			23		
8			16			24		

4. Is there a difference between the proportion of American men who are U.S. Senators and the proportion of U.S. Senators who are American men? If so, which proportion is greater?

5. Do you think it would be more common to see a toy advertisement in which a boy plays with a traditionally female toy or one in which a girl plays with a traditionally male toy?

6. Do you suspect that it is any more or less common for a physician to be a woman if she is in a certain age group? If so, in what age group would you expect to see the highest proportion of female physicians?

IN-CLASS ACTIVITIES

Activity 11-1: Penny Thoughts (*cont.*)

Reconsider the data collected in Topic 1 on students' political inclinations and opinions about whether the penny should be retained or abolished.

(a) For each student, *tally* which of the six possible category pairs he/she belongs to:

Category pair	Tally
Retain, liberal	
Retain, moderate	
Retain, conservative	
Abolish, liberal	
Abolish, moderate	
Abolish, conservative	

(b) Represent the *counts* of students falling into each of these category pairs in a two-way table. For example, the number that you place in the upper left cell of the table should be the number of students who classified themselves as liberal and also believe that the penny should be retained.

	Retain	Abolish
Liberal		
Moderate		
Conservative		

Activity 11-2: Age and Political Ideology

In a survey of adult Americans in 1986, people were asked to indicate their age and to categorize their political ideology. The results are summarized in the following table of counts:

	Liberal	Moderate	Conservative	Total
Under 30	83	140	73	296
30–49	119	280	161	560
Over 50	88	284	214	586
Total	290	704	448	1442

This table is called a *two-way table* since it classifies each person according to two variables. In particular, it is a 3 × 3 table; the first number represents the number of categories of the row variable (age), and the second number represents the number of categories of the column variable (political ideology). As an example of how to read the table, the upper-left entry indicates that of the 1442 respondents, 83 were under 30 years of age and regarded themselves as political liberals. Notice that the table also includes row and column totals.

(a) What proportion of the survey respondents were under age 30?

(b) What proportion of the survey respondents were between 30 and 50 years of age?

(c) What proportion of the survey respondents were over age 50?

You have calculated the *marginal distribution* of the age variable. When analyzing two-way tables, one typically starts by considering the marginal distribution of each of the variables by themselves before moving on to explore possible relationships between the two variables. You saw in Topic 1 that marginal distributions can be represented graphically in *bar graphs*. A bar graph illustrating the marginal distribution of the gender variable appears below:

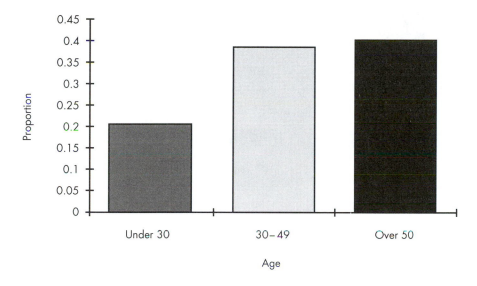

To study possible relationships between two categorical variables, one examines *conditional distributions*; i.e., distributions of one variable for *given* categories of the other variable.

(d) Restrict your attention (for the moment) to the respondents under 30 years of age. What proportion *of the young respondents* classify themselves as liberal?

(e) What proportion *of the young respondents* classify themselves as moderate?

(f) What proportion *of the young respondents* classify themselves as conservative?

One can proceed to calculate the conditional distributions of political ideology for middle-aged and older people. These turn out to be:

	Middle-aged	Older
Proportion liberal	0.2125	0.1502
Proportion moderate	0.5000	0.4846
Proportion conservative	0.2875	0.3652

Conditional distributions can be represented visually with *segmented bar graphs*. The rectangles in a segmented bar graph all have a height of 100%, but they contain *segments* whose length corresponds to the conditional proportions.

(g) Complete the segmented bar graph below by constructing the conditional distribution of political ideology among young people.

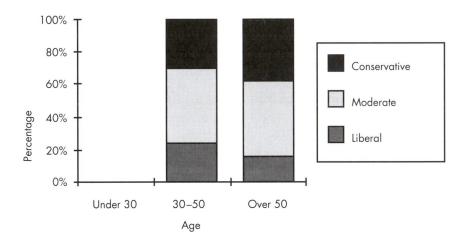

(h) Based on the calculations that you have performed and the display that you have created above, write a few sentences commenting on whether there seems to be any relationship between age and political ideology. In other words, does the distribution of political ideology seem to differ among the three age groups? If so, describe key features of the differences.

When dealing with conditional proportions, it is very important to keep straight which category is the one being conditioned on. For example, the proportion *of American males* who are U.S. Senators is very small, yet the proportion *of U.S. Senators* who are American males is very large.

Refer to the original table of counts to answer the following:

(i) What proportion of respondents under age 30 classified themselves as political moderates?

(j) What proportion of the political moderates were under 30 years of age?

(k) What proportion of the 1442 respondents identified themselves as being both under 30 years of age *and* political moderates?

Activity 11-3: Pregnancy, AZT, and HIV

In an experiment reported in the March 7, 1994 issue of *Newsweek*, 164 pregnant, HIV-positive women were randomly assigned to receive the drug AZT during pregnancy and 160 such women were randomly assigned to a control group which received a placebo ("sugar" pill). The following segmented bar graph displays the conditional distributions of the child's HIV status (positive or negative) for mothers who received AZT and for those who received a placebo.

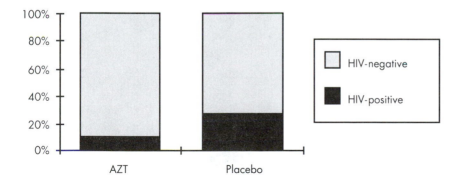

(a) Use the graph to estimate the proportion of AZT-receiving women who had HIV-positive babies and the proportion of placebo-receiving women who had HIV-positive babies.

The actual results of the experiment were that 13 of the mothers in the AZT group had babies who tested HIV-positive, compared to 40 HIV-positive babies in the placebo group.

(b) Use this information to calculate the proportions asked for in (a). Compare your calculations to your estimates based on the graph.

(c) The proportion of HIV-positive babies among placebo mothers is how many times greater than the proportion of HIV-positive babies among AZT-mothers?

(d) Comment on whether the difference between the two groups appears to be important. What conclusion would you draw from the experiment?

Activity 11-4: Hypothetical Hospital Recovery Rates

The following two-way table classifies hypothetical hospital patients according to the hospital that treated them and whether they survived or died:

	Survived	Died	Total
Hospital A	800	200	1000
Hospital B	900	100	1000

(a) Calculate the proportion of hospital A's patients who survived and the proportion of hospital B's patients who survived. Which hospital saved the higher percentage of its patients?

Suppose that when we further categorize each patient according to whether they were in good condition or poor condition prior to treatment we obtain the following two-way tables:

Good condition:

	Survived	Died	Total
Hospital A	590	10	600
Hospital B	870	30	900

Poor condition:

	Survived	Died	Total
Hospital A	210	190	400
Hospital B	30	70	100

(b) Convince yourself that when the "good"- and "poor"-condition patients are combined, the totals are indeed those given in the table above.

(c) Among those who were in *good* condition, compare the recovery rates for the two hospitals. Which hospital saved the greater percentage of its patients who had been in good condition?

(d) Among those who were in *poor* condition, compare the recovery rates for the two hospitals. Which hospital saved the greater percentage of its patients who had been in poor condition?

The phenomenon that you have just discovered is called *Simpson's paradox* (We think it's named for Lisa, but it could be for Bart or Homer), which refers to the fact that aggregate proportions can reverse the direction of the relationship seen in the individual pieces. In this case, hospital B has the higher recovery rate overall, yet hospital A has the higher recovery rate for each type of patient.

(e) Write a few sentences explaining (arguing from the data given) how it happens that hospital B has the higher recovery rate overall, yet hospital A has the higher recovery rate for each type of patient. (*Hints:* Do good or poor patients tend to survive more often? Does one type of hospital tend to treat most of one type of patient? Is there any connection here?)

(f) Which hospital would you rather go to if you were ill? Explain.

Activity 11-5: Hypothetical Employee Retention Predictions

Suppose that an organization is concerned about the number of its new employees who leave the company before they finish one year of work. In an effort to predict whether a new employee will leave or stay, they develop a standardized test and apply it to 100 new employees. After one year, they note what the test had predicted (stay or leave) and whether the employee actually stayed or left. They then compile the data into the following table:

	Actually stays	Actually leaves	Row total
Predicted to stay	63	12	75
Predicted to leave	21	4	25
Column total	84	16	100

(a) Of those employees predicted to stay, what proportion actually left?

(b) Of those employees predicted to leave, what proportion actually left?

(c) Is an employee predicted to stay any less likely to leave than an employee predicted to leave?

The segmented bar graph displaying the conditional distribution of employee retention between those predicted to stay and those predicted to leave follows:

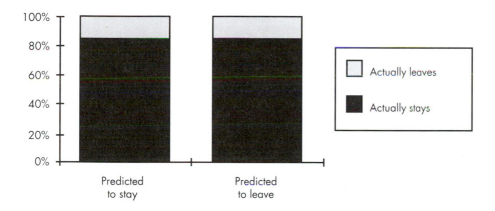

(d) Considering your answers to (a), (b), and (c) and this segmented bar graph, does the test provide any information at all about whether an employee will leave or stay?

Two categorical variables are said to be *independent* if the conditional distributions of one variable are identical for every category of the other variable. In this case the employee outcome is independent of the test prediction.

(e) Sketch what the segmented bar graph would look like if the test was *perfect* in its predictions.

(f) Sketch what the segmented bar graph would look like if the test was *very useful* but not quite perfect in its predictions.

HOMEWORK ACTIVITIES

Activity 11-6: Gender-Stereotypical Toy Advertising

To study whether toy advertisements tend to picture children with toys considered typical of their gender, researchers examined pictures of toys in a number of children's catalogs. For each picture, they recorded whether the child pictured was a boy or girl. (We will ignore ads in which boys and girls appeared together.) They also recorded whether the toy pictured was a traditional "male" toy (like a truck or a toy soldier) or a traditional "female" toy (like a doll or a kitchen set) or a "neutral" toy (like a puzzle or a toy phone). Their results are summarized in the following two-way table:

	Traditional "boy" toy	Traditional "girl" toy	Neutral gender toy
Boy child shown	59	2	36
Girl child shown	15	24	47

(a) Calculate the marginal totals for the table.
(b) What proportion of the ads showing *boy* children depicted traditionally male toys? traditionally female toys? neutral toys?

(c) Calculate the conditional distribution of toy types for ads showing *girl* children.

(d) Construct a segmented bar graph to display these conditional distributions.

(e) Based on the segmented bar graph, comment on whether the researchers' data seem to suggest that toy advertisers do indeed tend to present pictures of children with toys stereotypical of their gender.

Activity 11-7: Gender-Stereotypical Toy Advertising *(cont.)*

Reconsider the data concerning toy advertising presented in Activity 10-6. Let us refer to ads which show boys with traditionally "female" toys and ads which show girls with traditionally "male" toys as "crossover" ads.

(a) What proportion of the ads under consideration are "crossover" ads?

(b) What proportion *of the crossover ads* depict girls with traditionally male toys?

(c) What proportion *of the crossover ads* depict boys with traditionally female toys?

(d) When toy advertisers do defy gender stereotypes, in which direction does their defiance tend?

Activity 11-8: Female Senators

The following table classifies each U.S. Senator of 1994 according to his/her gender and political party:

	Male	Female	Row total
Republicans	42	2	44
Democrats	51	5	56
Column total	93	7	100

(a) What proportion of the Senators are women?

(b) What proportion of the Senators are Democrats?

(c) Are most Democratic Senators women? Support your answer with an appropriate calculation.

(d) Are most female Senators Democrats? Support your answer with an appropriate calculation.

Activity 11-9: *Jurassic Park* Popularity

Consider the data collected above in the "Preliminaries" section concerning whether or not students have read the novel and/or seen the movie *Jurassic Park*.

(a) Tabulate the data in a two-way table such as the following:

	Seen movie	Not seen movie
Read book		
Not read book		

(b) Calculate the conditional distributions of movie watching for those who have read the book and for those who have not read the book.
(c) Construct a segmented bar graph to display these conditional distributions.
(d) Write a brief paragraph summarizing whether the data seem to suggest an association between reading the book and seeing the movie.

Activity 11-10: Gender of Physicians *(cont.)*

The following data address the question of whether percentages of female physicians are changing with time. The table classifies physicians according to their gender and age group.

	Under 35	35–44	45–54	55–64
Male	93,287	153,921	110,790	80,288
Female	40,431	44,336	18,026	7,224

(a) For each age group, calculate the proportion of its physicians who are women.
(b) Construct a segmented bar graph to represent these conditional distributions.
(c) Comment on whether your analysis reveals any connection between gender and age group. Suggest an explanation for your finding.

Activity 11-11: Children's Living Arrangements

The following table classifies the living arrangements of American children (under 18 years of age) in 1993 according to their race/Hispanic origin and which parent(s) they live with.

	Both	Just Mom	Just Dad	Neither
White	40,842,340	9,017,140	2,121,680	1,060,840
Black	3,833,640	5,750,460	319,470	745,430
Hispanic	4,974,720	2,176,440	310,920	310,920

Analyze these data to address the issue of whether a relationship exists between race/Hispanic origin and parental living arrangements. Write a paragraph reporting your findings, supported by appropriate calculations and visual displays.

Activity 11-12: Civil War Generals

The following table categorizes each general who served in the Civil War according to his background before the war, and the Army (Union or Confederate) for which he served.

	Military	Law	Business	Politics	Agriculture	Other
Union	197	126	116	47	23	74
Confederate	127	129	55	24	42	48

Analyze these data to address the question of whether Union and Confederate generals tended to have different types of backgrounds. Perform calculations and construct displays to support your conclusions.

Activity 11-13: Berkeley Graduate Admissions

The University of California at Berkeley was charged with having discriminated against women in the graduate admissions process for the fall quarter of 1973. The table below identifies the number of acceptances and denials for both male and female applicants in each of the six largest graduate programs at the institution at that time:

	Men accepted	Men denied	Women accepted	Women denied
Program A	511	314	89	19
Program B	352	208	17	8
Program C	120	205	202	391
Program D	137	270	132	243
Program E	53	138	95	298
Program F	22	351	24	317
Total				

(a) Start by ignoring the program distinction, collapsing the data into a two-way table of gender by admission status. To do this, find the total number of men accepted and denied and the total number of women accepted and denied. Construct a table like the one below:

	Admitted	Denied	Total
Men			
Women			
Total			

(b) Consider for the moment just the *men* applicants. Of the men who applied to one of these programs, what proportion were admitted? Now consider the *women* applicants; what proportion of them were admitted? Do these proportions seem to support the claim that men were given preferential treatment in admissions decisions?

(c) To try to isolate the program or programs responsible for the mistreatment of women applicants, calculate the proportion of men and the proportion of women *within each program* who were admitted. Record your results in a table like the one below.

	Proportion of men admitted	Proportion of women admitted
Program A		
Program B		
Program C		
Program D		
Program E		
Program F		

(d) Does it seem as if any program is responsible for the large discrepancy between men and women in the overall proportions admitted?

(e) Reason from the data given to explain how it happened that men had a much higher rate of admission overall even though women had higher rates in most programs and no program favored men very strongly.

Activity 11-14: Baldness and Heart Disease

To investigate a possible relationship between heart disease and baldness, researchers asked a sample of 663 male heart patients to classify their degree of baldness on a 5-point scale. They also asked a control group (not suffering from heart disease) of 772 males to do the same baldness assessment. The results are summarized in the table:

	None	Little	Some	Much	Extreme
Heart disease	251	165	195	50	2
Control	331	221	185	34	1

(a) What proportion of these men identified themselves as having little or no baldness?

(b) Of those who had heart disease, what proportion claimed to have some, much, or extreme baldness?

(c) Of those who declared themselves as having little or no baldness, what proportion were in the control group?

(d) Construct a segmented bar graph to compare the distributions of baldness ratings between subjects with heart disease and those from the control group.

(e) Summarize your findings about whether a relationship seems to exist between heart disease and baldness.

(f) Even if a strong relationship exists between heart disease and baldness, does that necessarily mean that heart disease is *caused* by baldness? Explain your answer.

Activity 11-15: Softball Batting Averages

Construct your own hypothetical data to illustrate Simpson's paradox in the following context: Show that it is possible for one softball player (Amy) to have a higher percentage of hits than another (Barb) in the first half of the season and in the second half of the season and yet to have a lower percentage of hits for the season as a whole. I'll get you started: suppose that Amy has 100 at-bats in the first half of the season and 400 in the second half, and suppose that Barb has 400 at-bats in the first half and 100 in the second half. You are to make up how many hits each player had in each half of the season, so that the above statement holds. (The proportion of hits is the number of hits divided by the number of at-bats.)

	First half of season	Second half of season	Season as a whole
Amy's hits			
Amy's at-bats	100	400	500
Amy's proportion of hits			
Barb's hits			
Barb's at-bats	400	100	500
Barb's proportion of hits			

Activity 11-16: Employee Dismissals

Suppose that you are asked to investigate the practices of a company that has recently been forced to dismiss many of its employees. The company figures indicate that, of the 1000 men and 1000 women who worked there a month ago, 300 of the men and 200 of the women were dismissed. The company employs two types of employees - professional and clerical, so you ask to see the breakdown for each type. Even though the company dismissed a higher percentage of men than women, you know that it is possible (Simpson's paradox) for the percentage of women dismissed within each employee type to exceed that of the men within each type. The company representative does not believe this, however, so you need to construct a hypothetical example to convince him of the possibility. Do so, by constructing and filling in tables such as the following:

Overall (professional and clerical combined):

	Dismissed	Retained
Men	300	700
Women	200	800

Professional only: Clerical only:

	Dismissed	Retained
Men		
Women		

	Dismissed	Retained
Men		
Women		

Activity 11-17: Politics and Ice Cream

Suppose that 500 college students are asked to identify their preferences in political affiliation (Democrat, Republican, or Independent) and in ice cream (chocolate, vanilla, or strawberry). Fill in the following table in such a way that the variables *political affiliation* and *ice cream preference* turn out to be completely *independent.* In other words, the conditional distribution of ice cream preference should be the same for each political affiliation, and the conditional distribution of political affiliation should be the same for each ice cream flavor.

	Chocolate	Vanilla	Strawberry	Row total
Democrat	108			240
Republican		72	27	
Independent		32		80
Column total	225			500

Activity 11-18: Penny Thoughts (*cont.*)

Refer to the two-way table that you created in Activity 11-1 classifying students according to their opinion about the U.S. penny and their political leanings. Analyze this table to address the question of whether a relationship exists between these two variables. Write a paragraph summarizing your findings.

Activity 11-19: Variables of Personal Interest *(cont.)*

Think of a pair of categorical variables that you would be interested in exploring the relationship between. Describe the variables in as much detail as possible and indicate how you would present the data in a two-way table.

WRAP-UP

With this topic we have concluded our investigation of relationships between variables. This topic has differed from earlier ones in that it has dealt exclusively with categorical variables. The most important technique that this topic has covered has involved interpreting information presented in two-way tables. You have encountered the ideas of *marginal distributions* and *conditional distributions*, and you have learned to draw *bar graphs* and *segmented bar graphs* to display these distributions. Finally, you have discovered and explained the phenomenon known as *Simpson's paradox*, which raises interesting issues with regard to analyzing two-way tables.

These first two units have addressed exploratory analyses of data. In the next unit you will begin to study background ideas related to the general issue of drawing *inferences* from data. Specifically, you will take a figurative step backward in the data analysis process by considering issues related to the question of how to collect meaningful data in the first place. You will also begin to study ideas of *randomness* that lay the foundation for procedures of *statistical inference*.

Unit Three

Randomness

RANDOM SAMPLING

To this point in the text, you have been analyzing data using exploratory methods. With this topic you will take a conceptual step backward by considering questions of how to collect data in the first place. The practice of statistics begins not *after* the data have been collected but *prior* to their collection. You will find that utilizing proper data collection strategies is critical to being able to draw meaningful conclusions from the data once it has been collected and analyzed. You will also discover that *randomness* plays an important role in data collection.

- To understand the importance of proper data collection designs in order to draw meaningful conclusions from a study.
- To appreciate the fundamental distinctions between *population* and *sample* and between *parameter* and *statistic*.
- To recognize biased sampling methods and to be wary of conclusions drawn from studies which employ them.
- To discover the principle of *simple random sampling* and to be able to implement it using a *table of random digits*.
- To begin to develop an informal sense for some of the properties of randomness, particularly with regard to *sample size*.

PRELIMINARIES

1. What percentage of adult Americans do you think believes that Elvis Presley is still alive?

2. Who won the 1936 U.S. Presidential election? Who lost that election?

3. If Ann takes a random sample of 5 Senators and Barb takes a random sample of 25 Senators, who is more likely to come close to the actual percentage breakdown of Democrats/Republicans in her sample?

4. Is she (your answer to the previous question) *guaranteed* to come closer to this percentage breakdown?

5. Which would have more days on which 60% or more of its births were boys: a small hospital which has 20 births per day or a large one which has 100 births per day?

IN-CLASS ACTIVITIES

Activity 12-1: Elvis Presley and Alf Landon

On the twelfth anniversary of the (alleged) death of Elvis Presley, a Dallas record company sponsored a national call-in survey. Listeners of over 1000 radio stations were asked to call a 1-900 number (at a charge of $2.50) to voice an opinion concerning whether or not Elvis was really dead. It turned out that 56% of the callers felt that Elvis was still alive.

(a) Do you think that 56% is an accurate reflection of beliefs of *all* American adults on this issue? If not, identify some of the flaws in the sampling method.

In 1936, *Literary Digest* magazine conducted the most extensive (to that date) public opinion poll in history. They mailed out questionnaires to over 10 million people whose names and addresses they had obtained from phone books and vehicle registration lists. More than 2.4 million people responded, with 57% indicating that they would vote for Republican Alf Landon in the upcoming Presidential election. (Incumbent Democrat Franklin Roosevelt won the election, carrying 63% of the popular vote.)

(b) Offer an explanation as to how the *Literary Digest*'s prediction could have been so much in error. In particular, comment on why its sampling method made it vulnerable to overestimating support for the Republican candidate.

These examples have (at least) two things in common. First, the goal in each case was to learn something about a very large group of people (all American adults, all American registered voters) by studying a *portion* of that group. That is the essential idea of *sampling:* to learn about the whole by studying a part.

Two extremely important terms related to the idea of sampling that we will use throughout the text are *population* and *sample*. In the technical sense with which we use these terms, *population* means the *entire* group of people or objects about which information is desired, while *sample* refers to a (typically small) *part* of the population that is actually examined to gain information about the population.

(c) Identify the population of interest and the sample actually used to study that population in the Elvis and *Literary Digest* examples.

Another thing that the two examples have in common is that both illustrate a very poor job of sampling; i.e., of selecting the sample from the population. In neither case could one accurately infer anything about the population of interest from the sample results. This is because the sampling methods used were *biased*. A sampling procedure is said to be *biased* if it tends *systematically* to overrepresent certain segments of the population and systematically to underrepresent others.

These examples also indicate some common problems that produce biased samples. Both are *convenience samples* to some extent since they both reached those people most readily accessible. Another problem is *voluntary response*, which refers to samples collected in such a way that members of the population decide for themselves whether or not to participate in the sample. The related problem of *nonresponse* can arise even if an unbiased sample of the population is contacted.

Activity 12-2: Sampling U.S. Senators

In order to avoid biased samples, it seems fair and reasonable to give each and every member of the population the same chance of being selected for the sample. In other words, the sample should be selected so that *every* possible sample has an equal chance of being the sample ultimately selected. Such a sampling design is called *simple random sampling.*

While the principle of simple random sampling is probably clear, it is by no means simple to implement. One thought is to rely on physical mixing: write the names on pieces of paper, throw them into a hat, mix them thoroughly, and draw them out one at a time until the sample is full. Unfortunately, this method is fraught with the potential for hidden biases, such as different sizes of pieces of paper and insufficient mixing.

A better alternative for selecting a simple random sample (hereafter to be abbreviated SRS) is to use a calculator-generated *table of random digits.* Such a table is constructed so that each position is equally likely to be occupied by any one of the digits 0, 1, 2, 3, 4, 5, 6, 7, 8, 9 and so that the occupant of any one position has no impact on the occupant of any other position. A table of random digits follows:

17139	27838	19139	82031	46143	93922	32001	05378	42457	94248
20875	29387	32682	86235	35805	66529	00886	25875	40156	92636
34568	95648	79767	16307	71133	15714	44142	44293	19195	30569
11169	41277	01417	34656	80207	33362	71878	31767	04056	52582
15529	30766	70264	86253	07179	24757	57502	51033	16551	66731
33241	87844	41420	10084	55529	68560	50069	50652	76104	42086
83594	48720	96632	39724	50318	91370	68016	06222	26806	86726
01727	52832	80950	27135	14110	92292	17049	60257	01638	04460
86595	21694	79570	74409	95087	75424	57042	27349	16229	06930
65723	85441	37191	75134	12845	67868	51500	97761	35448	56096
82322	37910	35485	19640	07689	31027	40657	14875	07695	92569
06062	40703	69318	95070	01541	52249	56515	59058	34509	35791
54400	22150	56558	75286	07303	40560	57856	22009	67712	19435
80649	90250	62962	66253	93288	01838	68388	55481	00336	19271
70749	78066	09117	62350	58972	80778	46458	83677	16125	89106
50395	30219	03068	54030	49295	48985	016247	28818	83101	18172
48993	89450	04987	02781	37935	76222	93595	20942	90911	57643
77447	34009	20728	88785	81212	08214	93926	66687	58252	18674
24862	18501	22362	37319	33201	88294	55814	67443	77285	36229
87445	26886	66782	89931	29751	08485	49910	83844	56013	26596

Consider the members of the U.S. Senate of 1994 as the population of interest. These Senators are listed below, along with their sex, party, state, and years of service (as of 1994) in the Senate. Notice that each has been assigned a two-digit identification number.

ID#	Name	Sex	Party	State	Years
01	Akaka	m	Dem	Hawaii	4
02	Baucus	m	Dem	Montana	16
03	Bennett	m	Rep	Utah	1
04	Biden	m	Dem	Delaware	21
05	Bingaman	m	Dem	New Mexico	11
06	Bond	m	Rep	Missouri	7
07	Boren	m	Dem	Oklahoma	15
08	Boxer	f	Dem	California	1
09	Bradley	m	Dem	New Jersey	15
10	Breaux	m	Dem	Louisiana	7
11	Brown	m	Rep	Colorado	3
12	Bryan	m	Dem	Nevada	5
13	Bumpers	m	Dem	Arkansas	19
14	Burns	m	Rep	Montana	5
15	Byrd	m	Dem	West Virginia	35
16	Campbell	m	Dem	Colorado	1
17	Chafee	m	Rep	Rhode Island	18
18	Coats	m	Rep	Indiana	5
19	Cochran	m	Rep	Mississippi	16

ID#	Name	Sex	Party	State	Years
20	Cohen	m	Rep	Maine	15
21	Conrad	m	Dem	North Dakota	7
22	Coverdell	m	Rep	Georgia	1
23	Craig	m	Rep	Idaho	3
24	D'Amato	m	Rep	New York	13
25	Danforth	m	Rep	Missouri	18
26	Daschle	m	Dem	South Dakota	7
27	DeConcini	m	Dem	Arizona	17
28	Dodd	m	Dem	Connecticut	13
29	Dole	m	Rep	Kansas	25
30	Domenici	m	Rep	New Mexico	21
31	Dorgan	m	Dem	North Dakota	1
32	Durenberger	m	Rep	Minnesota	16
33	Exon	m	Dem	Nebraska	15
34	Faircloth	m	Rep	North Carolina	1
35	Feingold	m	Dem	Wisconsin	1
36	Feinstein	f	Dem	California	1
37	Ford	m	Dem	Kentucky	20
38	Glenn	m	Dem	Ohio	20
39	Gorton	m	Rep	Washington	13
40	Graham	m	Dem	Florida	7
41	Gramm	m	Rep	Texas	9
42	Grassley	m	Rep	Iowa	13
43	Gregg	m	Rep	New Hampshire	1
44	Harkin	m	Dem	Iowa	9
45	Hatch	m	Rep	Utah	17
46	Hatfield	m	Rep	Oregon	27
47	Heflin	m	Dem	Alabama	15
48	Helms	m	Rep	North Carolina	21
49	Hollings	m	Dem	South Carolina	28
50	Hutchison	f	Rep	Texas	1
51	Inouye	m	Dem	Hawaii	31
52	Jeffords	m	Rep	Vermont	5
53	Johnston	m	Dem	Louisiana	22
54	Kassebaum	f	Rep	Kansas	16
55	Kempthorne	m	Rep	Idaho	1
56	Kennedy	m	Dem	Massachusetts	32
57	Kerry, J	m	Dem	Massachusetts	9
58	Kerry, R	m	Dem	Nebraska	5
59	Kohl	m	Dem	Wisconsin	5
60	Lautenberg	m	Dem	New Jersey	12
61	Leahy	m	Dem	Vermont	19
62	Levin	m	Dem	Michigan	15
63	Lieberman	m	Dem	Connecticut	5
64	Lott	m	Rep	Mississippi	5
65	Lugar	m	Rep	Indiana	17
66	Mack	m	Rep	Florida	5
67	Matthews	m	Dem	Tennessee	1

ID#	Name	Sex	Party	State	Years
68	McCain	m	Rep	Arizona	7
69	McConnell	m	Rep	Kentucky	9
70	Metzenbaum	m	Dem	Ohio	18
71	Mikulski	f	Dem	Maryland	7
72	Mitchell	m	Dem	Maine	14
73	Moseley-Braun	f	Dem	Illinois	1
74	Moynihan	m	Dem	New York	17
75	Murkowski	m	Rep	Alaska	13
76	Murray	f	Dem	Washington	1
77	Nickles	m	Rep	Oklahoma	13
78	Nunn	m	Dem	Georgia	22
79	Packwood	m	Rep	Oregon	25
80	Pell	m	Dem	Rhode Island	33
81	Pressler	m	Rep	South Dakota	15
82	Pryor	m	Dem	Arkansas	15
83	Reid	m	Dem	Nevada	7
84	Riegle	m	Dem	Michigan	18
85	Robb	m	Dem	Virginia	5
86	Rockefeller	m	Dem	West Virginia	9
87	Roth	m	Rep	Delaware	23
88	Sarbanes	m	Dem	Maryland	17
89	Sasser	m	Dem	Tennessee	17
90	Shelby	m	Dem	Alabama	7
91	Simon	m	Dem	Illinois	9
92	Simpson	m	Rep	Wyoming	15
93	Smith	m	Rep	New Hampshire	3
94	Specter	m	Rep	Pennsylvania	13
95	Stevens	m	Rep	Alaska	26
96	Thurmond	m	Rep	South Carolina	38
97	Wallop	m	Rep	Wyoming	17
98	Warner	m	Rep	Virginia	15
99	Wellstone	m	Dem	Minnesota	16
00	Wofford	m	Dem	Pennsylvania	3

Some characteristics of this population are:

Males:	93
Females:	7

Democrats:	56
Republicans:	44

	Mean	Std. dev.	Min	Q_1	Median	Q_3	Max
Years of service	12.54	8.72	1	5	13	17	38

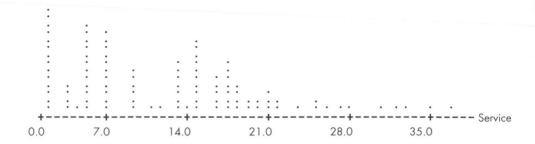

(a) For each of the four variables that are recorded about the Senators, identify whether it is a categorical or measurement variable. If it is categorical, specify whether it is also binary.

Sex:

Party:

State:

Years of service:

(b) Use the table of random digits to select a simple random sample of 10 U.S. Senators. Do this by entering the table at *any* point and reading off the first 10 two-digit numbers that you happen across. (If you happen to get repeats, keep going until you have ten different two-digit numbers.) Record the names and other information of the Senators corresponding to those ID numbers:

	ID	Senator	Sex	Party	State	Years
1						
2						
3						
4						
5						
6						
7						
8						
9						
10						

(c) Enter in the table below the numbers of men, women, Democrats, and Republicans in your sample.

Males:	
Females:	

Democrats:	
Republicans:	

(d) What is your home state? Is there a senator from your home state in the sample?

(e) Create (by hand) a dotplot of the years of service in your sample. Then use the calculator to compute summary statistics for the distribution of years of service and record your findings in the table below.

Years of service	Mean	Std. dev.	Min	Q_1	Median	Q_3	Max

(f) Does the proportional breakdown of men/women in your sample equal that in the entire population of Senators? How about the breakdown of Democrats/Republicans? Does the mean years of service in your sample equal that of the population?

(g) If your answer to any part of question (f) is "no," does that mean that your sampling *method* is biased like the *Literary Digest*'s was? Explain.

(h) Describe specifically how you would have to alter this procedure to use the table of random digits to select an SRS of 10 from the population of the 435 members of the U.S. House of Representatives.

Although the ideas of population and sample are crucial in statistics, a closely related distinction concerns quantities called parameters and statistics. A *parameter* is a numerical characteristic of a *population*, while a *statistic* is a numerical characteristic of a *sample*. (To help you keep this straight, notice that *population* and *parameter* start with the same letter, as do *sample* and *statistic*.)

We will be very careful to use different symbols to denote parameters and statistics. For example, we use the following symbols to denote proportions, means, and standard deviations:

	(Population) Parameter	(Sample) Statistic
Proportion:	θ	$\hat{p}$
Mean:	μ	$\bar{x}$
Standard deviation:	σ	s

(Note: $\hat{p}$ is read "p-hat" and $\bar{x}$ is read "x-bar.")

(i) Identify each of the following as a parameter or a statistic, indicate the symbol used to denote it, and specify its value in this context.

- the proportion of men in the entire 1994 Senate

- the proportion of Democrats among your 10 Senators

- the mean years of service among your 10 Senators

- the standard deviation of the years of service in the 1994 Senate

Activity 12-3: Sampling U.S. Senators (*cont.*)

As you may have already guessed, the calculator can generate simple random samples more quickly than you can.

(a) In order to take simple random samples of U.S. Senators you need to use a program named SENATORS.83p. Download this program into your calculator.

(b) Use SENATORS.83p to generate a simple random sample of 10 U.S. Senators. The program generates a random list of ten different numbers corresponding to ID numbers of the Senators. Once you are through with the list of ID numbers, press ENTER to view both the number of Democrats and the mean years of service for this sample. In the table below, record the sample *proportion* of Democrats in the sample and the sample mean of the years of service of those Senators in your sample. Repeat this a total of ten times.

Sample	1	2	3	4	5	6	7	8	9	10
Prop. Dem.										
Mean years										

(c) Did you get the same sample proportion of Democrats in each of your 10 samples? Did you get the same sample mean years of service in each of your 10 samples?

This simple question illustrates a very important statistical property known as *sampling variability*: the value of sample quantities vary from sample to sample.

(d) Create (by hand) a dotplot of your sample proportions of Democrats. (The value marked with the arrow is .56, the population proportion of Democrats.)

(e) Use the calculator to compute the mean and standard deviation of your sample proportions of Democrats.

Mean:

Standard deviation:

(f) Now use the SENATORS program to select ten samples of *forty* Senators in each sample (it would be wise to work with a partner to complete this exercise). Again record the sample *proportion* of Democrats in each sample and the sample mean of the years of service of those Senators in each sample. Create (by hand) a dotplot of your sample proportions of Democrats on the axis below.

Sample	1	2	3	4	5	6	7	8	9	10
Prop. Dem.										
Mean years										

(g) Again use the calculator to compute the mean and standard deviation of your sample proportions of Democrats.

Mean:

Standard deviation:

(h) Comparing the two plots that you have produced, in which case (samples of size 10 or samples of size 40) is the *variability* among sample proportions of Democrats greater?

(i) In which case (samples of size 10 or samples of size 40) is the result of a *single* sample more likely to be close to matching the truth about the population?

(j) If you were to draw a very large number (thousands or millions) of simple random samples of size 10 and to record the sample proportion of Democrats in each sample, and if you were to then calculate the mean of the sample proportions, what number would you expect this mean to be very close to?

(k) Would your answer to (j) be different if each sample was of size 40 Senators rather than 10?

This exercise has illustrated that the value of a sample statistic (the sample proportion of Democrats, in this case) varies from sample to sample if one repeatedly takes simple random samples from the population of interest. This simple idea of sampling variability is in fact a very important and powerful one that lays the foundation for most techniques of statistical inference.

The last few questions have also introduced you to the crucial role that the *sample size* plays in planning sample surveys and interpreting their results. The larger the size of a random sample, the more likely the sample is to accurately reflect the entire population.

Two caveats are in order, however. First, one still gets the occasional "unlucky" sample whose results are not close to the population even with large sample sizes. Second, the sample size means little if the sampling method is not random. Remember that the *Literary Digest* had a huge sample of 2.4 million people, yet their results were not close to the truth about the population.

HOMEWORK ACTIVITIES

Activity 12-4: Sampling U.S. Senators (*cont.*)

Reconsider the samples of U.S. Senators that you produced (with the help of the calculator) above.

(a) Create (by hand) dotplots of the distributions of *sample means* of years of service. Put the two different cases (sample size of 10 and sample size of 40) on different axes but on the same scale (such as the following).

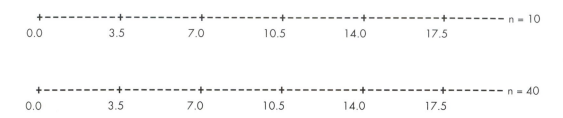

(b) Use the calculator to compute the means and standard deviations of the 10 sample means for each sample size. Record these values.
(c) With which sample size (10 or 40) is the variability in sample means greater?
(d) Are the means of these sample means roughly the same with both sample sizes?

Activity 12-5: Emotional Support

In the mid-1980's, Shere Hite undertook a study of women's attitudes toward relationships, love, and sex by distributing 100,000 questionnaires through women's groups. Of the 4500 women who returned the questionnaires, 96% said that they give more emotional support than they receive from their husbands or boyfriends.

An ABC News/*Washington Post* poll surveyed a random sample of 767 women, finding that 44% claimed to give more emotional support than they receive.

(a) Comment on whether Hite's sampling method is likely to be biased in a particular direction. Specifically, do you think the 96% figure overestimates or underestimates the truth about the population of all American women?
(b) Which poll surveyed the larger number of women?
(c) Which poll's results do you think are more representative of the truth about the population of all American women? Explain.

Activity 12-6: Alternative Medicine

In a spring 1994 issue, *Self* magazine reported that 84% of its readers who responded to a mail-in poll indicated that they had used a form of alternative medicine (e.g., acupuncture, homeopathy, herbal remedies). Comment on whether this sample result is representative of the truth concerning the population of all adult Americans. Do you suspect that the sampling method has biased the result? If so, is the sample result likely to overestimate or underestimate the proportion of all adult Americans who have used alternative medicine? Explain your answers.

Activity 12-7: Courtroom Cameras

An article appearing in the October 4, 1994, issue of *The Harrisburg Evening-News* reported that Judge Lance Ito (who is trying the O.J. Simpson murder case) had received 812 letters from around the country on the subject of whether to ban cameras from the courtroom. Of these 812 letters, 800 expressed the opinion that cameras should be banned.

(a) What proportion of this sample supports a ban on cameras in the courtroom? Is this number a parameter or a statistic?

(b) Do you think that this sample represents well the population of all American adults? Comment on the sampling method.

Activity 12-8: Boy and Girl Births

Suppose that a city has two hospitals. Hospital A has about 100 births per day, while Hospital B has only about 20 births per day. Assume that each birth is equally likely to be a girl or a boy. Suppose that for one year you count the number of days on which a hospital has 60% or more of that day's births turn out to be boys. Which hospital would you expect to have more such days? Explain your reasoning. (Try to figure out how this question relates to the issue of random sampling and sample size.)

Activity 12-9: Parameters Versus Statistics

(a) Suppose that you are interested in the population of all students at your school and that you are using the students enrolled in your course as a (nonrandom) sample. Identify each of the following as a parameter or a statistic, and indicate the symbol used to denote it:
- the proportion of the college's students who participate in inter-collegiate athletics
- the proportion of students in your course who participate in inter-collegiate athletics
- the mean grade point average for the students in your course
- the standard deviation of the grade point averages for all students at your school

(b) Suppose that you are interested in the population of all college students in the United States and that you are using students at your college as a (nonrandom) sample. Identify each of the following as a parameter or a statistic, and indicate the symbol used to denote it.
- the proportion of your college's students who have a car on campus
- the proportion of all U.S. college students who have a car on campus
- the mean financial aid amount being received this academic year for all U.S. college students
- the standard deviation of the financial aid amounts being received this academic year for all students at your college

Activity 12-10: Nonsampling Sources of Bias

(a) Suppose that simple random samples of adult Americans are asked to complete a survey describing their attitudes toward the death penalty. Suppose that one group is asked, "Do you believe that the U.S. judicial system should have the right to call for executions?" while another group is asked, "Do you believe that the death penalty should be an option in cases of horrific murder?". Would you anticipate that the proportions of "yes" responses might differ between these two groups? Explain.

(b) Suppose that simple random samples of students on your campus are questioned about a proposed policy to ban smoking in all campus build-ings. If one group is interviewed by a person wearing a T-shirt and jeans and smoking a cigarette while another group is interviewed by a non-smoker wearing a business suit, would you expect that the proportions declaring agreement with the policy might differ between these two groups? Explain.

(c) Suppose that an interviewer knocks on doors in a suburban community and asks the person who answers whether he/she is married. If the person is married, the interviewer proceeds to ask, "Have you ever engaged in extramarital sex?" Would you expect the proportion of "yes" responses to be close to the actual proportion of married people in the community who have engaged in extramarital sex? Explain.

(d) Suppose that simple random samples of adult Americans are asked whether or not they approve of the President's handling of foreign policy. If one group is questioned prior to a nationally televised speech by the President on his/her foreign policy and another is questioned immediately after the speech, would you be surprised if the proportions of people expressing approval differed between these two groups? Explain.

(e) List four sources of bias that can affect sample survey results even if the sampling procedure used is indeed a random one. Base your list on the preceding four questions.

Activity 12-11: Survey of Personal Interest

Find a newspaper, magazine, or televised account of the results of a recent survey of interest to you. Write a description of the survey in which you identify the variable(s) involved, the population, the sample, and the sampling method. Also comment on whether the sampling method seems to have been random and, if not, whether it seems to be biased in a certain direction.

WRAP-UP

The material covered in this topic differs from earlier material in that you are beginning to consider issues related to how to collect data in the first place. This topic has introduced four terms that are central to formal statistical inference: *population* and *sample*, *parameter* and *statistic*.

One of the key ideas to take away from this topic is that a poor method of collecting data can lead to misleading (if not completely meaningless) conclusions. Another fundamental idea is that of *random sampling* as a means of selecting a sample that will (most likely) be representative of the population that one is interested in.

At this stage you have also begun to investigate (informally) properties of randomness. For instance, you have explored the notion of *sampling variability*, particularly noting the effect of the *sample size*.

In the next topic you will continue to explore properties of sampling variability and will, in fact, discover that this type of variability follows a very specific long-term pattern known as a *sampling distribution*.

SAMPLING DISTRIBUTIONS I: CONFIDENCE

You have been studying a strategy for data collection which involves the use of randomization. At first glance, it might seem that this deliberate introduction of uncertainty would only compound the problem of drawing reliable conclusions from the data collected. On the contrary, you will discover that the randomness actually produces very predictable long-term patterns of variation. The knowledge of these long-term patterns of variation forms the basis for using sample results to draw inferences about populations.

- To gain an understanding of the fundamental concept of *sampling variability*.
- To discover and understand the concept of *sampling distributions* as representing the long-term pattern of variation of a statistic under repeated sampling.
- To explore the sampling distribution of a sample proportion through actual experiments and calculator simulations.
- To discover the effect of sample size on the sampling distribution of a sample proportion.
- To investigate the principle of *statistical confidence* as it relates to estimating a population parameter based on a sample statistic.

PRELIMINARIES

1. Which color of Reese's Pieces candies do you think is the most common: orange, brown, or yellow?

2. Take a guess concerning the proportion of Reese's Pieces candies which have that color.

3. If each student in your class takes a random sample of 25 Reese's Pieces candies, would you expect every student to obtain the same number of orange candies?

4. What percentage of the popular vote in the 1992 U.S. Presidential election went to Bill Clinton? to George Bush? to Ross Perot?

5. Take a guess concerning the proportion of American households that include a pet cat.

6. How would you respond to the question, "Do you think the United States is in a moral and spiritual decline?"

7. In a national survey conducted in June of 1994, what proportion of American adults do you think would have responded in the affirmative to the question about moral decline?

8. In Activities 13-1 and 13-2 you will be asked to study distributions of proportions of orange candies from random samples of Reese's Pieces. To complete Activity 13-2 you need the calculator to simulate taking 500 random samples of size 25 and 75 of Reese's Pieces. The name of the

program needed to do this simulation is SIMSAMP.83p. Download this program into your calculator.

Choose a partner (if you do not have one already) and run SIM-SAMP.83p. One calculator should have 25 entered for sample size while the other should have 75. Both calculators need 500 entered for the number of samples and .45 entered for the true population proportion (we are assuming that 45% of the population is orange). This program places all of the data into a list named PROP.

The simulation for the samples of size 25 will take approximately 8 minutes to run while the simulation for the samples of size 75 will take approximately 22 minutes. The results of these simulations will be used in Activity 13-2. (Warning: This program requires a significant amount of memory. You should delete any lists from your calculator that are no longer necessary).

IN-CLASS ACTIVITIES

Recall from the previous topic that a population is comprised of the entire group of people or objects of interest to an investigator, while a sample refers to the part of the population that he or she actually studies. Also remember that a parameter is a numerical characteristic of a population and that a statistic is a numerical characteristic of a sample.

In certain contexts a *population* can also refer to a *process* (such as flipping a coin or manufacturing a candy bar) which in principle can be repeated indefinitely. With this interpretation of population, a *sample* is a specific collection of process outcomes.

Activity 13-1: Colors of Reese's Pieces Candies

Consider the *population* of the Reese's Pieces candies manufactured by Hershey. Suppose that you want to learn about the distribution of colors of these candies but that you can only afford to take a *sample* of 25 candies.

(a) Take a random sample of 25 candies and record the number and proportion of each color in your sample.

	Orange	Yellow	Brown
Number			
Proportion			

(b) Is the proportion of orange candies among the 25 that you selected a *parameter* or a *statistic*? What symbol did we introduce in Topic 11 to denote it?

(c) Is the proportion of orange candies manufactured by Hershey's process a parameter or a statistic? What symbol represents it?

(d) Do you *know* the value of the proportion of orange candies manufactured by Hershey?

(e) Do you know the value of the proportion of orange candies among the 25 that you selected?

These simple questions point out the important fact that one typically knows (or can easily calculate) the value of a sample statistic, but only in very rare cases does one know the value of a population parameter. Indeed, a primary goal of sampling is to *estimate* the value of the parameter based on the statistic.

(f) Do you suspect that every student in the class obtained the same proportion of orange candies in his or her sample? By what term did we refer to this phenomenon in the previous topic?

(g) Use the axis below to construct a dotplot of the sample proportions of orange candies obtained by the students in the class.

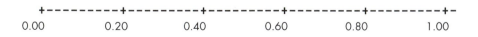

```
+---------+---------+---------+---------+---------+-
0.00      0.20      0.40      0.60      0.80      1.00
```

(h) *Did* everyone obtain the same number of orange candies in their samples?

(i) If every student was to estimate the population proportion of orange candies by the proportion of orange candies in his or her sample, would everyone arrive at the same estimate?

(j) Based on what you learned in the previous topic about random sampling and having the benefit of seeing the sample results of the entire class, take a guess concerning the population proportion of orange candies.

(k) Again assuming that each student had access only to her or his sample, would most estimates be reasonably close to the true parameter value? Would some estimates be way off? Explain.

(l) Remembering what you learned in the last topic, in what way would the dotplot have looked different if each student had taken a sample of *10* candies instead of 25?

(m) Remembering what you learned in the last topic, in what way would the dotplot have looked different if each student had taken a sample of *75* candies instead of 25?

Our class results suggest that even though sample values vary depending on which sample you happen to pick, there seems to be a *pattern* to this variation. We need more samples to investigate this pattern more thoroughly, however. Since it is time-consuming (and possibly fattening) to *literally* sample candies, we will use the calculator to *simulate* the process.

Activity 13-2: Simulating Reese's Pieces

To perform these simulations we needed to suppose that we knew the actual value of the parameter. For the simulations started in the preliminaries section we supposed that 45% of the population is orange.

(a) Use the results of the calculator simulation of drawing 500 samples of 25 candies each. (Pretend that this is really 500 students, each taking 25 candies and counting the number of orange ones). Then look at a display of the sample *proportions* of orange obtained.

(b) Do you notice any *pattern* in the way that the resulting 500 sample proportions vary? Explain.

(c) Use the calculator to compute the mean and standard deviation of these sample proportions.

Mean:

Standard Deviation:

(d) Roughly speaking, are there more sample proportions *close* to the population proportion (which, you will recall, is .45) than there are *far* from it?

(e) Let us quantify the previous question. Use the TRACE button of the calculator to count how many of the 500 sample proportions are within ±.10 of .45 (i.e., between .35 and .55). Then repeat for within ±.20 and for within ±.30. Record the results below:

	Number of the 500 sample proportions	Percentage of these sample proportions
Within ± .10 of .45		
Within ± .20 of .45		
Within ± .30 of .45		

(f) Forget for the moment that you have designated that the population proportion of orange candies be .45. Suppose that each of the 500 imaginary students was to estimate the population proportion of orange candies by going a distance of .20 on either side of her/his sample proportion. What percentage of the 500 students would capture the actual population proportion (.45) within this interval?

(g) Still forgetting that you actually know the population proportion of orange candies to be .45, suppose that you were one of those 500 imaginary students. Would you have any way of knowing *definitively* whether your sample proportion was within .20 of the population proportion? Would you be reasonably "confident" that your sample proportion was within .20 of the population proportion? Explain.

The pattern displayed by the variation of the sample proportions from sample to sample is called the *sampling distribution* of the sample proportion. Even though the sample proportion of orange candies varies from sample to sample, there is a recognizable long-term pattern to that variation. Thus, while one cannot use a sample proportion to estimate a population proportion *exactly*, one can be reasonably *confident* that the population proportion is within a certain distance of the sample proportion.

This "distance" depends primarily on how confident one wants to be and on the size of the sample.

(h) Use the results from your calculator simulation of drawing 500 samples of 75 candies each (these samples are three times larger than the ones you gathered in class and simulated earlier). Look at a display of the sample *proportions* and write down their mean and standard deviation.

Mean:

Standard deviation:

(i) How has the sampling distribution changed from when the sample size was only 25 candies?

(j) Use the $\boxed{\text{TRACE}}$ button of the calculator to count how many of these 500 sample proportions are within $\pm.10$ of .45. Record this number and the percentage below.

(k) How do the percentages of sample proportions falling within $\pm.10$ of .45 compare between sample sizes of 25 and 75?

(l) In general, is a sample proportion more likely to be close to the population proportion with a larger sample size or with a smaller sample size?

Since these sample proportions follow a distribution which is symmetric and mound-shaped, the *empirical rule* (of Topic 5) establishes that about 95% of the sample proportions fall within two standard deviations of the mean.

(m) Remembering that you found the mean and standard deviation of these sample proportions in (h), double the standard deviation. Then subtract this value from the mean and also add this value to the mean. Record the results.

(n) Use the calculator to count how many of the sample proportions fall within this interval. What percentage of the 500 sample proportions is this? Is this percentage close to the 95% predicted by the empirical rule?

(o) If each of the 500 imaginary students would subtract this value (twice the standard deviation) from her or his sample proportion and also add this value to her or his sample proportion, about what percentage of the students' intervals would contain the actual population proportion of .45?

This activity reveals that if one wants to be about 95% confident of capturing the population proportion within a certain distance of one's sample proportion, that "distance" should be about twice the standard deviation of the sampling distribution of sample proportions.

One need not use simulations to determine how much sample proportions vary. A theoretical result establishes that if one repeatedly takes simple random samples of size n from a population for which the *true* proportion possessing the attribute of interest is θ, then the sampling distribution of the sample proportion $\hat{p}$ has mean equal to θ and standard deviation equal to $\sqrt{\theta(1 - \theta)/n}$.

(p) Continuing to assume that the population proportion of orange candies is $\theta = .45$, what does the theoretical result say about the mean and standard deviation of the sampling distribution of sample proportions when the sample consists of $n = 25$ candies? Do these values come close to your simulated results in (c) above?

Mean:

Standard deviation:

(q) Repeat the previous question for a sample size of $n = 75$. Compare your theoretical answers to your simulated results in (h).

Mean:

Standard deviation:

HOMEWORK ACTIVITIES

Activity 13-3: Parameters vs. Statistics (*cont.*)

Identify each of the following as a parameter or a statistic, and also indicate the symbol used to denote it. In some cases you may have to form your own conclusion as to what the population of interest is.

  (a) the proportion of students in your class who have visited Europe
  (b) the proportion of all students at your school who have visited Europe
  (c) the mean number of states visited by the students in your class
  (d) the mean number of states visited by all students in your school
  (e) the standard deviation of the years of service of the 100 U.S. Senators
  (f) the standard deviation of the years of service of the 10 U.S. Senators who comprised the random sample you generated using the table of random digits
  (g) the proportion of American voters from the 1992 election who voted for Bill Clinton
  (h) the proportion of people at the next party you attend who voted for George Bush in the 1992 election
  (i) the proportion of American voters who voted for Ross Perot in the 1992 election
  (j) the mean number of cats in the households of the faculty members of your school
  (k) the mean number of cats among all American households
  (l) the proportion of "heads" in 100 coin flips
  (m) the mean weight of 20 bags of potato chips

Activity 13-4: Presidential Votes

In the 1992 U.S. Presidential election, Bill Clinton received 43% of the popular vote, compared to 38% for George Bush and 19% for Ross Perot. (Each of these figures has been rounded.) Suppose that we take a simple random sample of 100 voters from that election and ask them for whom they voted.

(a) Would it necessarily be the case that these 100 voters would include 43 Clinton supporters, 38 Bush supporters, and 19 Perot supporters?

Now suppose that you were to repeatedly take SRS's of 100 voters.

(b) Would you find the same sample proportion of Clinton supporters each time?
(c) According to the theoretical result presented above, what would be the standard deviation of the sampling distribution of the sample proportion of Clinton voters?
(d) According to the empirical rule, about 95% of your samples would find the sample proportion of Clinton voters to be between what two values?
(e) Repeat (c) for the sample proportion of Bush supporters.
(f) Repeat (c) for the sample proportion of Perot supporters.
(g) With which candidate would you see the most variation in the sample proportions who voted for him? With which candidate would you see the least variation?

Activity 13-5: Presidential Votes *(cont.)*

Bill Clinton received 43% of the votes cast in the 1992 Presidential election. Suppose that you were to take an SRS of size n from the population of all votes cast in that election.

(a) Use the theoretical result presented above to calculate the standard deviation of the sampling distribution of these sample proportions for each of the following values of n: 50, 100, 200, 400, 500, 800, 100, 1600, 2000. (You may use the calculator.)
(b) Construct (by hand) a scatterplot of these standard deviations vs. the sample size n.
(c) By how many times does the sample size have to increase in order to cut this standard deviation in half?

Activity 13-6: Presidential Votes *(cont.)*

Let θ represent the proportion of votes received by a certain candidate in an election. Suppose that you repeatedly take SRS's of 100 voters and calculate the sample proportion who voted for the candidate.

(a) Use the theoretical result presented above to calculate the standard deviation of the sampling distribution of these sample proportions for each of the following values of θ: 0, .1, .2, .3, .4, .5, .6, .7, .8, .9, 1. (You may use the calculator.)
(b) Construct (by hand) a scatterplot of these standard deviations vs. the population proportion θ.
(c) Which value of θ produces the most variability in sample proportions?
(d) Which values of θ produce the least variability in sample proportions? Explain in a sentence or two what happens in these cases.

Activity 13-7: American Moral Decline

In a survey conducted by *Newsweek* on June 2–3, 1994, 748 adults were asked, "Do you think the United States is in a moral and spiritual decline?" Of those who responded, 76% said "yes."

(a) Is this 76% a parameter or a statistic? Explain.
(b) Do you think that *exactly* 76% of *all* American adults would have answered "yes" to the question?

(c) Based on this sample survey result, do you think that the proportion of *all* American adults who would have answered "yes" to the question is closer to 76% or to 36%?

(d) Roughly how many percentage points would you have to go on either side of 76% to be about 95% confident that the interval would include the actual proportion of *all* American adults who would have answered "yes"?

Activity 13-8: Cat Households

Suppose that you want to estimate the proportion of all households in your hometown which have a pet cat. (Let us call this proportion θ.) Suppose further that you take an SRS of 200 households, while your polling competitor takes an SRS of only 50 households.

(a) Can you be certain that the sample proportion of cat households in your sample will be closer to θ than your competitor's sample proportion?

(b) Do you have a better chance than your competitor of obtaining a sample proportion of cat households which falls within ±.05 of the actual value θ?

(c) Suppose for the moment that $\theta = .25$. Use the theoretical result to determine the standard deviation of the sampling distribution of sample proportions when the sample size is $n = 200$. Then do the same calculation for a sample size of $n = 50$. Which sample size produces the smaller standard deviation? How many times smaller is it than the other one?

(d) Repeat (c) but supposing that $\theta = .45$. Comment on the extent to which this change in the population proportion affects your conclusions.

Activity 13-9: Cat Households *(cont.)*

Suppose now that the actual proportion of households that have a pet cat (among the entire population of households in your hometown) is $\theta = .25$.

(a) Use SIMSAMP.83p to simulate the random selection of 100 samples, each sample containing 200 households. Enter 200 for the sample size, 100 for the number of samples, and .25 for the true population proportion. This will take several minutes to complete so make sure that your batteries are strong. The program places all of the proportions into a list named PROP. How many of these 100 samples produce a sample proportion within $\pm.05$ of the population proportion?

(b) Use the calculator to compute the five-number summary of the distribution of these sample proportions. Record the results.

(c) Again use SIMSAMP.83p to simulate the random selection of 100 samples, this time with each sample containing only 50 households. How many of these 100 samples produce a sample proportion within $\pm.05$ of the population proportion?

(d) Again use the calculator to compute the five-number summary of the distribution of these sample proportions, and record the results.

(e) Construct (by hand) boxplots of the distributions of these sample proportions. The boxplots should be on the same scale, but they need not be modified ones.

Activity 13-10: Boy and Girl Births (*cont.*)

Reconsider Activity 12-8. Use SIMSAMP.83p to simulate 365 days of births in each hospital. Continue to assume that each birth is equally likely to be a boy or a girl. Have the calculator keep track of the proportion of boys born on each day. For Hospital A you will need to enter 100 for the sample size, 365 for the number of samples, and 0.5 for the true population proportion. For Hospital B you will need to enter 20 for the sample size, 365 for the number of samples, and 0.5 for the true population proportion. This simulation will take several minutes to complete. On how many of these 365 days does hospital A have 60% or more of its births turn out to be boys? What about hospital B?

WRAP-UP

This topic has emphasized the fundamental distinction between a *parameter* and a *statistic;* you have also encountered the important notion of a population as a *process*. Most importantly, you have explored the obvious (but crucial) concept of *sampling variability* and learned that this variability displays a very definite pattern in the long run. This pattern is known as the *sampling distribution* of the statistic.

You have investigated properties of sampling distributions in the context of sample *proportions*. You have discovered that larger sample sizes produce less variation among sample proportions. A theoretical result has provided you a means of measuring that variation.

In addition, you have begun to explore how sampling distributions relate to the important idea of statistical *confidence:* that one can have a certain amount of confidence that the observed value of a sample statistic falls within a certain distance of the unknown value of a population parameter.

The next topic continues your study of sampling distributions, focusing on how they relate to the concept of *statistical significance.*

■

Topic 14:

SAMPLING DISTRIBUTIONS II: SIGNIFICANCE

You have been studying properties of randomness, specifically investigating how sample statistics vary from sample to sample under repeated random sampling from a population. This topic asks you to continue to develop your understanding of *sampling distributions*. The contexts in which you study these sampling distributions will introduce the very important idea of *statistical significance*.

- To continue to develop an understanding of sampling distributions.
- To discover the connection between sampling distributions and the issue of *statistical significance*.
- To investigate the role of sample size in questions of statistical significance.
- To begin to understand the reasoning process which underlies statistical tests of significance.

PRELIMINARIES

1. If one-third of the widgets produced by a manufacturer are defective and if you inspect random batches of 15 widgets coming off the assembly line, about how many defectives would you expect to see in a batch?

2. If different inspectors take different samples of 15 widgets from the assembly line, would you expect each of them to find the same number of defectives?

3. Would you be very surprised to see four or fewer defectives in a batch of 15 coming from a widget assembly line which produces one-third defectives in the long run?

4. Would you be very surprised to see two or fewer defectives in a batch of 15 coming from a widget assembly line which produces one-third defectives in the long run?

5. Would you be very surprised to see no defectives in a batch of 15 coming from a widget assembly line which produces one-third defectives in the long run?

6. If someone tells you that they have both good news and bad news to share with you, would you generally prefer to hear the good news first or the bad news first?

7. Pool the results for the class and record the numbers who prefer to hear good news first and who prefer to hear bad news first.

8. In Activities 14-1 and 14-2 you will be asked to study the proportion of defective widgets in a manufacturing process. To obtain a better understanding of this problem, you will need the calculator to perform a simulation of the random selection of 1,000 batches of 15 widgets. A program named WIDGET.83p has been designed to perform this simulation. Download this program into your calculator.

Choose a partner (if you do not have one already). One calculator should run WIDGET.83p by entering 1,000 for the number of batches while the other calculator should be available for computations. WIDGET.83p places all of the data into a list named WIDGT. The results of this simulation will be used in Activity 14-1 (l) − (o) and Activity 14-2. The simulation will take approximately 20 minutes.

IN-CLASS ACTIVITIES

Activity 14-1: Widget Manufacturing

Suppose that a widget manufacturer finds that *one-third* of the widgets coming off its assembly line are defective. Suppose further that the company's engineers have designed some modifications for the manufacturing process that they believe will decrease this proportion of defective widgets. They plan to sample and inspect a *batch* of 15 widgets to test this belief.

(a) Is the number one-third a parameter or a statistic? Explain. What symbol have we used to represent such a number?

(b) Suppose that the engineers sample one batch of 15 widgets and find 4 defectives. What is the sample proportion of defectives? (Also indicate the symbol used to represent this number.) Is the sample proportion of defectives less than one-third?

(c) Is it *possible* to have found 4 or fewer defectives in the sample batch even if the modifications had *no effect* on the process; i.e., even if the population proportion of defectives is still one-third? Explain.

(d) If the engineers were to sample 100 batches of widgets, would you expect each batch to have the same number of defectives? By what term do we refer to this phenomenon?

This activity sets the stage for you to study the concept of *statistical significance*. The idea is to explore the sampling distribution of a statistic, investigating how often an observed sample result would occur just by chance. Roughly speaking, a sample result is said to be *statistically significant* if it is unlikely to occur simply due to sampling variability alone.

(e) Suppose for now that the modifications to the process have *no effect* whatsoever; in other words, assume that the population proportion of defectives is still one-third. Use a fair six-sided die to simulate the random selection of a batch of widgets. Roll the die 15 times. If the die lands on 1 or 2, regard that widget as defective; when it lands on 3, 4, 5, or 6, consider that widget non-defective. Fill in the following table representing the batch, labeling defectives with D and non-defectives with N.

Widget	1	2	3	4	5	6	7	8	9	10	11	12	13	14	15
Die result															
Defective?															

(f) How many of the 15 widgets in your simulated batch are defective? What *proportion* of the batch is defective?

(g) Repeat this simulation until you have filled a total of five batches. (In other words, you will roll the die a total of 75 times, 15 times for each of 5 batches.) For each of your five simulated batches, record the number and proportion of defective widgets in the batch. (Round the proportion to three decimal places.)

Batch	1	2	3	4	5
Defective widgets					
Proportion defective					

(h) Did you obtain the same number of defectives in each sample?

(i) Combine your simulation results with those of the rest of the class by creating a dotplot on the axis below of the number of defective widgets in each simulated batch.

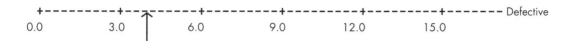

```
+---------+-------+---------+---------+---------+-------- Defective
0.0       3.0  ↑  6.0       9.0      12.0      15.0
```

(j) How many and what proportion of these simulated batches produced four or fewer defectives?

(k) Based on this simulation, would you say that it is *very unlikely* for the process to produce a batch with four or fewer defectives when the population proportion of defectives is one-third?

(l) Refer to the results of the calculator's simulation of the random selection of 1,000 batches of widgets, still assuming that one-third of the population is defective. The data in WIDGT are set up so that the first entry is the number of simulated batches with zero defects, the second entry is the number of simulated batches with one defective widget, and so forth. Have the calculator compute the number and proportion of defectives in each batch. Tally the simulated sample results in the table below.

Defective widgets	0	1	2	3	4	5	6	7
no. of simulated batches								
Defective widgets	8	9	10	11	12	13	14	15
no. of simulated batches								

(m) How many and what proportion of these 1000 simulated batches contain four or fewer defectives?

(n) Based on this more extensive simulation, would you say that it is *very unlikely* for the process to produce a batch with four or fewer defectives when the population proportion of defectives is one-third?

(o) Suppose again that the engineers do not know whether or not the modifications have improved the production process, so they sample a batch of 15 widgets and find four defectives. Does this finding provide strong evidence that they have improved the process? Explain.

Activity 14-2: Widget Manufacturing (*cont.*)

Consider again the situation of Activity 14-1.

(a) Now suppose that the engineers find just two defective widgets in their sample batch. About how often in the long run would the process produce such an extreme result (two or fewer defectives) if the modifications did not improve the process (i.e., if the population proportion of defectives were still one-third). Base your answer on the 1000 simulated batches that you generated above.

(b) Would finding two defectives in a sample batch provide strong evidence that the modification had in fact improved the process by lessening the proportion of defectives produced? Explain.

(c) Repeat (a) and (b) if the engineers find no defectives in the sample batch.

Statistical significance addresses the question of whether sample results are unlikely enough to have occurred by chance for one to conclude that another explanation is warranted. In this widget manufacturing example, it is extremely unlikely for a batch to have *no* defectives if the modifications do not improve the process. If the engineers make the changes and then find that a sample batch has no defectives, then logic dictates that they believe one of two possibilities:

(a) that the process has not been improved and a very unusual sample of widgets has been selected by chance; or

(b) that the modifications have improved the process and reduced the population proportion of defectives to less than one-third.

Either of these possibilities could be true, but the sample evidence (no defectives) is extremely unlikely under possibility (a). Notice that while this reasoning process does provide strong evidence for (b) in this case, it cannot *definitively* rule out either possibility.

On the other hand, finding four defectives in the sample batch provides very little evidence that the modifications improve the process, because it is not terribly unusual to observe four or fewer defectives when the process is running at its usual rate of one-third defectives.

This type of reasoning underlies widely used procedures of statistical inference known as *tests of significance*. We will study these procedures in detail in future topics.

Activity 14-3: ESP Testing

A common test for extra-sensory perception (ESP) asks subjects to identify which of four shapes (star, circle, wave, or square) appears on a card unseen by the subject. Consider a test of $n = 30$ cards. If a person does not have ESP and is just guessing, he/she should therefore get 25% right in the long run. In other words, the proportion of correct responses that the guessing subject would make in the long run would be $\theta = .25$. The following histogram displays the results of 10,000 simulated tests in which the subject guessed randomly on each card:

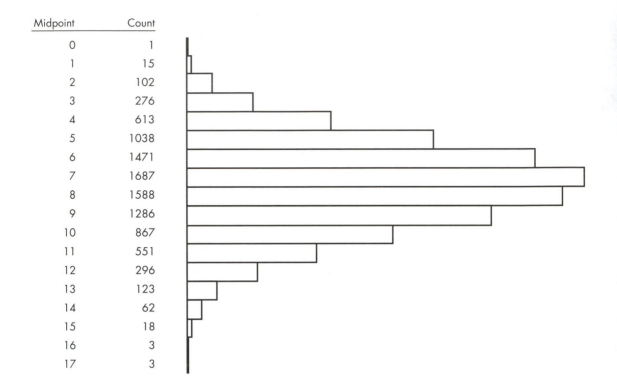

Midpoint	Count
0	1
1	15
2	102
3	276
4	613
5	1038
6	1471
7	1687
8	1588
9	1286
10	867
11	551
12	296
13	123
14	62
15	18
16	3
17	3

(a) In how many of these 10,000 simulated tests does the guessing subject identify *exactly* 25% of the 30 cards correctly?

(b) In how many of these 10,000 simulated tests does the sample proportion of correct responses exceed .25? What percentage of the 10,000 tests is this?

(c) How surprising would it be for a subject to get nine or more correct if he/she were just guessing? Refer to the results of the simulated tests in your answer.

(d) How surprising would it be for a subject to get 12 or more correct if he or she were just guessing? Again refer to the results of the simulated tests in your answer.

(e) How surprising would it be for a subject to get 15 or more correct if he or she were just guessing? Once again refer to the results of the simulated tests.

(f) Suppose that a particular subject gets 17 correct in a test. How convinced would you be that she actually possesses the ability to get more than 25% correct in the long run? Explain, basing your argument on the results of the simulated tests.

HOMEWORK ACTIVITIES

Activity 14-4: ESP Testing (*cont.*)

Refer back to the 10,000 simulated ESP tests presented in Activity 14-3.

(a) How many of the 30 cards would a subject have to identify correctly so that less than 10% of all guessing subjects do that well?

(b) Repeat (a) with 5% instead of 10%.

(c) Repeat (a) with 1% instead of 10%.

Activity 14-5: Jury Selection

Suppose that 40% of those eligible for jury duty in a certain community are senior citizens over 65 years of age.

(a) Is the number 40% a parameter or a statistic? Explain.
(b) If 12-person juries are selected as simple random samples from the population of those eligible for jury service, would *every* jury contain 5 senior citizens (which is as close as one can get to 40% of 12)?
(c) Is it possible that a randomly selected jury from this population would contain as few as two senior citizens? Could a randomly chosen jury contain as many as 10 senior citizens? Would it be possible for a randomly selected jury to contain no senior citizens?

The following histogram (at top of next page) represents the number of senior citizens in 10,000 randomly selected 12-person juries from this population with 40% senior citizens.

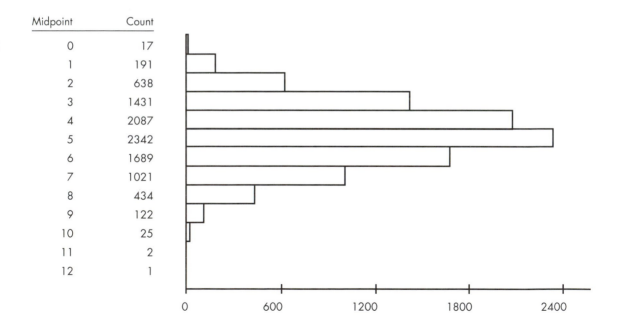

Midpoint	Count
0	17
1	191
2	638
3	1431
4	2087
5	2342
6	1689
7	1021
8	434
9	122
10	25
11	2
12	1

(d) Suppose that a particular defendant in a jury trial has reason to believe that senior citizens would be more sympathetic to her than younger jurors would be. Suppose that the jury selected to hear her case contains only two senior citizens. What percentage of the 10,000 randomly chosen juries contain so few (or fewer) senior citizens?
(e) Based on the simulation results, would you say that the defendant in (d) has strong evidence that the jury system actually selected senior citizens for jury duty at a rate less than 40%? Explain.

Activity 14-6: Penny Thoughts *(cont.)*

Reconsider the data collected in Topic 1 concerning students' attitudes about whether the United States should retain or abolish the penny as a coin of currency. Suppose that 50% of the population of students at this school favor retaining the penny. Use SIMSAMP.83p to simulate 999 random samples of size n, where n is the number of students in this class who responded to the penny question. Have the calculator produce visual displays of the distribution of the 999 sample proportions. Then comment on where the sample proportion for this class falls in the distribution. Would the sample result we obtained in class be very surprising if in fact 50% of the population of students at this school favor retaining the penny?

Activity 14-7: Good News/Bad News

Perform the analysis asked for in Activity 14-6 for the issue of whether students generally prefer to hear good news first or bad news first. Use the data collected in the "Preliminaries" section.

Activity 14-8: Home Court Advantage

Suppose that you want to investigate the notion of a "home court advantage," that the home team wins more than half of all games in the National Basketball Association. You decide to keep track of a sample of 30 NBA games.

(a) If there is no home court advantage (or disadvantage), what proportion of all NBA games would be won by the home team? Is this number a parameter or a statistic?

(b) Use SIMSAMP.83p to simulate 999 repetitions of 30 NBA games, assuming that the home team actually wins 50% of all games (i.e., assuming no home court advantage). This simulation may take about 45 minutes to be completed. Look at a histogram of the distribution of sample games won by the home team.

(c) Of the 30 NBA games played on January 3–7, 1992, 18 were won by the home team. What is the sample proportion of games won by the home team on these days?

(d) In how many and what proportion of your 999 simulated samples did the home team win 18 or more of the games, even though your simulations assumed *no* home court advantage?

(e) Based on the results of your simulation, would you say that it would be *very unlikely* for a home team to win 18 or more of a sample of 30 games even if there were no home court advantage? Explain.

(f) Again based on the results of your simulation, would you say that the sample evidence of 18 home wins in 30 games provides strong evidence of a home court advantage? Explain.

(g) Would you be convinced of a home court advantage if 24 of the 30 games sampled had been won by the home team? Explain.

WRAP-UP

You have continued your study of sampling distributions with this topic, concentrating your attention on issues of *statistical significance*. You have learned that the question of statistical significance relates to how often an observed sample result would occur by chance alone. Through physical (rolling dice) and calculator simulations, you have investigated some questions related to this important concept.

To this point you have relied on calculator simulations to get a rough sense of how often a certain sample result would occur by chance in the long run. In the next topic you will learn how to calculate such *probabilities* more precisely through the use of *normal distributions*.

Topic 15:

NORMAL DISTRIBUTIONS

You have been studying properties of randomness in general and sampling variability in particular. Through calculator simulations, you have discovered that sample statistics vary according to a very definite pattern. In this topic you will investigate mathematical models known as *normal distributions* which describe this pattern of variation very accurately. You will learn how to use normal distributions to calculate *probabilities* of interest in a variety of contexts.

- To become familiar with the idea of using *normal curves* as mathematical models for approximating certain distributions.
- To develop the ability to perform calculations related to *standard normal distributions* with the use of a *table of standard normal probabilities*.
- To discover how to use a table of standard normal probabilities to perform calculations pertaining to any normal distribution.

1. What is Forrest Gump's IQ?

2. Suppose that Professors Fisher and Savage assign A's to students scoring above 90 and F's to those scoring below 60 on the final exam. Suppose further that the distribution of scores on Professor Fisher's final has mean 74 and standard deviation 7, while the distribution of scores on Professor Savage's final has mean 78 and standard deviation 18. Which professor would you expect to assign more A's?

3. Would you expect Professor Fisher or Professor Savage to assign more F's?

4. From which one, Professor Fisher or Professor Savage, would you personally rather take a course (all other factors being equal)?

5. How many days do you think a human pregnancy lasts on the average?

6. About what proportion of human pregnancies would you guess last less than 8 months? more than 10 months?

In-Class Activities

Activity 15-1: Placement Scores and Reese's Pieces

The following dotplot of the data from Activity 5-3 reveals the distribution of scores on the mathematics placement exam taken by Dickinson College freshmen in 1992:

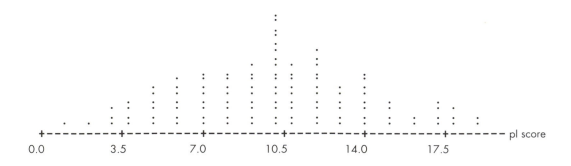

The next dotplot displays the sample proportions of orange Reese's Pieces from 500 simulated samples of size 25 (each dot represents 5 points):

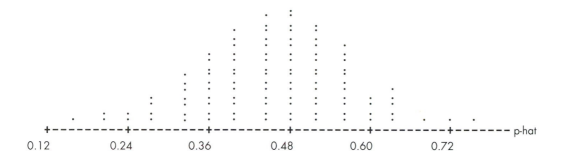

(a) What similarities do you notice about the shapes of these distributions?

(b) Draw a sketch of a smooth curve ("smooth" meaning having no breaks or jagged edges) that seems to approximate the general shape apparent in the two dotplots above.

Data that display the general shape seen in the examples above occur very frequently. Theoretical mathematical models used to approximate such distributions are called *normal distributions*. Notice the use of the plural here, for there is not just *one* normal curve; there is a whole *family* of them. Each member of the family of normal distributions shares three distinguishing characteristics, however:

• Every normal distribution is symmetric.
• Every normal distribution has a single peak at its center.
• Every normal distribution follows a bell-shaped curve.

Each member of the family of normal distributions is identified by two things: its *mean* and its *standard deviation*. The mean μ determines where its center is; the peak of a normal curve is its mean, which is also its point of symmetry. The standard deviation σ indicates how spread out the distribution is. The distance between the mean μ and the points where the curvature changes (in other words, where you would have to start turning the wheel if you were driving a car along the normal curve) is equal to the standard deviation σ. The sketch displays this relationship:

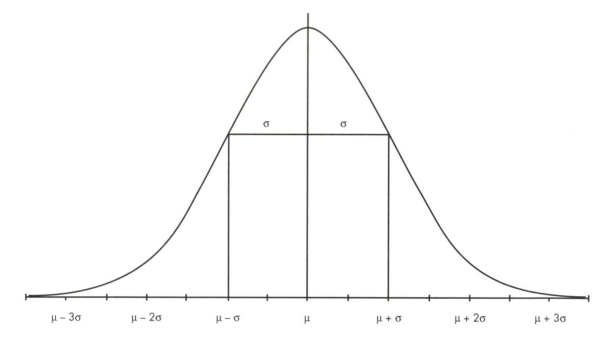

$\mu - 3\sigma$ $\mu - 2\sigma$ $\mu - \sigma$ μ $\mu + \sigma$ $\mu + 2\sigma$ $\mu + 3\sigma$

(c) The sketch below contains three normal curves; think of them as approximating the distribution of exam scores for three different classes. One (call it A) has a mean of 70 and a standard deviation of 5; another (call it B) has a mean of 70 and a standard deviation of 10; the third (call it C) has a mean of 50 and a standard deviation of 10. Identify which is which by labeling each curve with its appropriate letter.

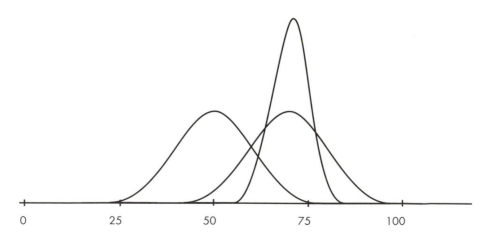

0   25   50   75   100

One can use normal curves to calculate (approximately, at least) the proportion of a population's observations that fall in a given interval of values. This proportion is also known as the *probability* that the value of a particular member of the population will fall in the given interval. Geometrically, this proportion (or probability) is equivalent to finding the *area* under the curve over the given interval. The *total* area under the curve is always equal to 1 (or 100%).

For instance, if scores on a certain exam are normally distributed with mean 70 and standard deviation 10, one could (theoretically) calculate the proportion of scores between 80 and 90 by finding the area under that normal curve between 80 and 90:

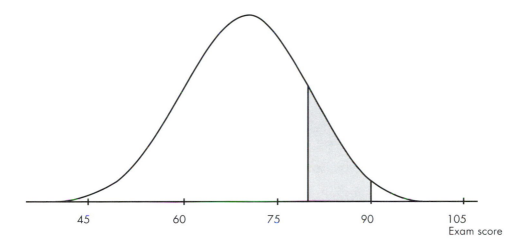

The question remains as to how to find such areas. For one particular member of the family of normal curves, these areas have been extensively tabulated. This member of the family is the one with mean $\mu = 0$ and standard deviation $\sigma = 1$; it is known as the *standard normal distribution* and is almost always denoted by the letter Z. Table I in the back of the book lists areas under the standard normal curve.

Activity 15-2: Standard Normal Calculations

Consider a sketch of the standard normal distribution:

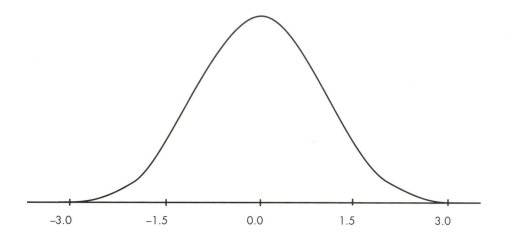

This activity will give you practice using the table of standard normal probabilities (a skill that you will need throughout the text). To use the table efficiently, you need to make use of three facts:

1. The table reports the area under the standard normal curve to the *left* of a point.
2. The *total* area under the curve is 1, so the area under the curve to the *right* of a point can be found by subtracting the tabled value for the point from 1.
3. The standard normal curve is *symmetric*, so the area to the *right* of a point equals the area to the *left* of that point's *negative*.

It is very important to keep notation straight in your mind at this point. We continue to let Z denote the standard normal curve, and we will use Pr to denote "proportion" or "probability." For example, the proportion of a standard normal distribution which falls less than 1.67 would be written as $\Pr(Z < 1.67)$.

Practice using the standard normal table to find the following probabilities. In each case, sketch the area that you are looking for under the standard normal curve drawn. (It is *always* a good idea to draw such a sketch, partly to remind yourself that probabilities correspond to areas, partly because the sketch can help you to figure out how to use the table correctly, and partly because it allows you to check visually whether your answer seems to be a reasonable one.)

(a) $\Pr(Z < 0.68)$

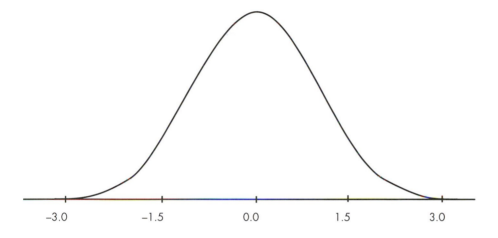

(b) $\Pr(Z \le 0.68)$ [Hint: Is the *area* of this region any different than the area of the previous one?]

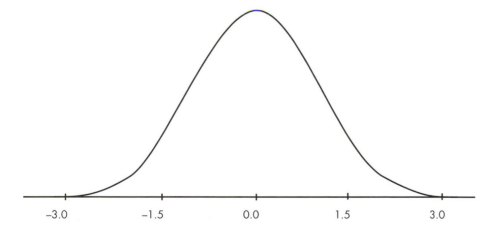

(c) $\Pr(Z > 0.68)$ [Try to think of two different ways to find this.]

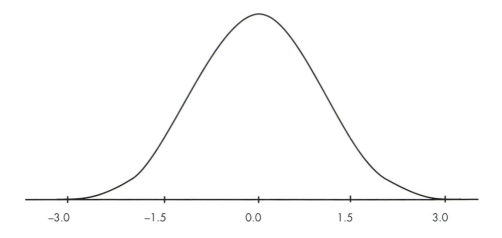

(d) $\Pr(Z < -1.38)$

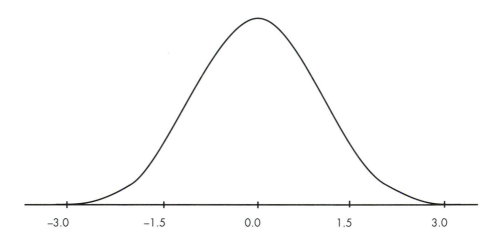

(e) Pr$(-1.38 < Z < 0.68)$ [Figure out how to find this by drawing a sketch and using your answers to earlier questions.]

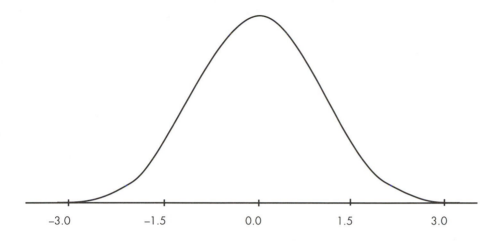

(f) Pr$(Z < -3.81)$ [Due to the limitations of your table, you will only be able to give an approximate answer here. Use the information in the sketch.]

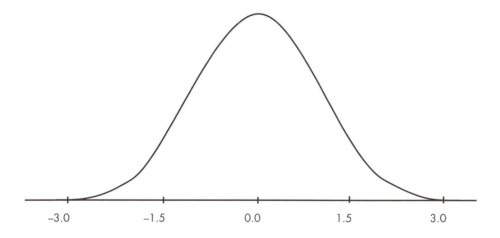

(g) Find the value k such that $\Pr(Z < k) = .8997$. [Draw a sketch and think about it. You will have to use the table in "reverse."] Your answer here is the 90th *percentile* of the standard normal distribution (technically, the 89.97th percentile), because 90% of the distribution falls below it.

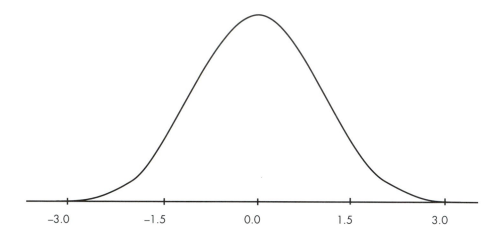

(h) Find k such that $\Pr(Z > k) = .0630$.

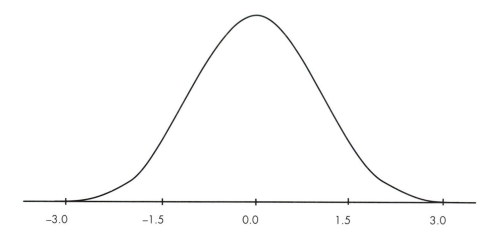

Activity 15-3: IQ Scores

At this point you have mastered all that you need to know concerning the use of the standard normal table. Unfortunately, most interesting normally distributed variables (SAT scores, pregnancy durations, etc.) do not have *standard* normal distributions.

Fortunately, the technique of *standardization* allows you to use simple algebra to convert *any* normal distribution into a standard normal distribution (at which point you can use the standard normal table). As you discovered when you first encountered the standard deviation in Topic 4, one *standardizes* by subtracting the mean μ of the distribution and then dividing by the standard deviation σ of the distribution. In symbols, if we let X denote any normal distribution (and continue to let Z denote the standard normal distribution), then standardization says that

$$Z = \frac{X - \mu}{\sigma}.$$

For an application of standardization, let X denote the variable of the IQ scores of the students in a certain college. Suppose that these IQ scores are known to follow a normal distribution with mean $\mu = 112$ and standard deviation $\sigma = 12$. A sketch of this distribution is shown below.

(a) The area corresponding to the proportion of IQs that are less than 100 has been shaded on the sketch below. Based on this visual impression (and remembering that the *total* area under any normal curve equals 1, or 100%), make a guess as to the proportion of IQ scores less than 100.

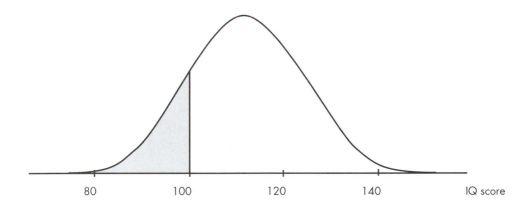

(b) Standardize the value of 100 by subtracting the mean of the distribution from 100 and then dividing that result by the standard deviation of the distribution.

(c) On the *standard* normal curve drawn below, shade in the area to the left of the standardized value that you just calculated. [The area that you are looking for with regard to the IQs is equivalent to this area under the standard normal curve, which you know how to find from the table.]

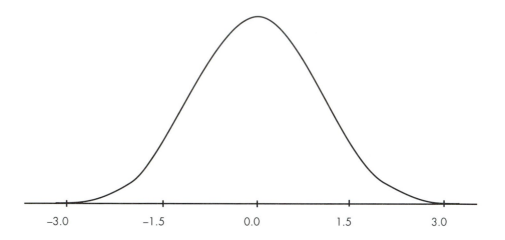

(d) Look up this standardized value in the table of standard normal probabilities to find the proportion of IQs that are less than 100.

Notationally, the best way to represent the calculation you just performed is as follows:

$$Pr(X < 100) = Pr(Z < -1.00) = .1587.$$

Use the technique of standardization and the standard normal table to answer the following. (Remember that it is always a good idea to draw yourself a sketch.)

(e) Find the proportion of these undergraduates having IQs greater than 130.

(f) Find the proportion of these undergraduates having IQs between 110 and 130.

(g) With his IQ of 75, Forrest Gump would have a higher IQ than what percentage of these undergraduates?

(h) Determine how high one's IQ must be to be in the top 1% of all IQs at this college. [This is similar to parts (g) and (h) of Activity 14-2, but you also need to consider standardization.]

(i) A program named PROB.83p uses the built-in features of your calculator to enable you to determine the area of a region under a normal curve. Download PROB.83p into your calculator and use this program to check your calculations in (e), (f), and (g).

Homework Activities

Activity 15-4: Normal Curves

For each of the following normal curves, identify (as accurately as you can from the graph) the mean μ and standard deviation σ of the distribution.

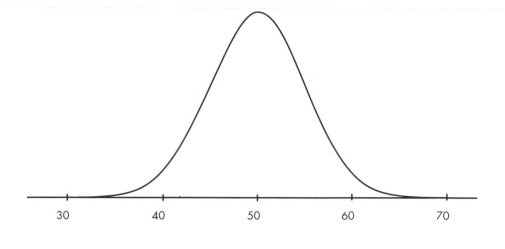

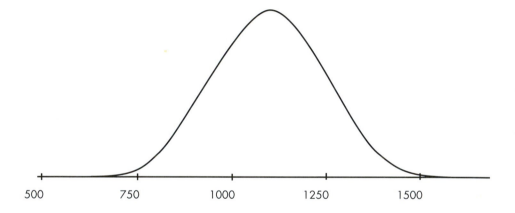

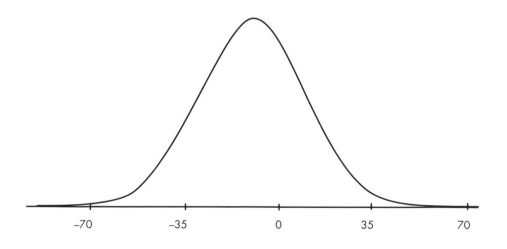

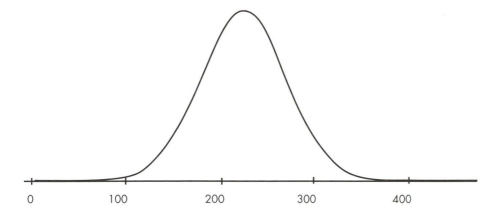

Activity 15-5: Pregnancy Durations

The distribution of the duration of human pregnancies (i.e., the number of days between conception and birth) has been found to be approximately normal with mean $\mu = 266$ and standard deviation $\sigma = 16$. Determine the proportion of all pregnancies that last:

(a) less than 244 days (which is about 8 months)
(b) more than 275 days (which is about 9 months)
(c) over 300 days
(d) between 260 and 280 days

Activity 15-6: Professors' Grades

Suppose that you are deciding whether to take Professor Fisher or Professor Savage next semester. You happen to know that each professor gives A's to those scoring above 90 on the final exam and F's to those scoring below 60. You also happen to know that the distribution of scores on Professor Fisher's final is approximately normal with mean 74 and standard deviation 7 and that the distribution of scores on Professor Savage's final is approximately normal with mean 78 and standard deviation 18.

(a) Produce a rough sketch of both teachers' grade distributions (on the same scale).
(b) Which professor gives the higher proportion of A's? Show the appropriate calculations to support your answer.
(c) Which professor gives the higher proportion of F's? Show the appropriate calculations to support your answer.

(d) Suppose that Professor DeGroot has a policy of giving A's to the top 10% of the scores on his final, regardless of the actual scores. If the distribution of scores on his final turns out to be normal with mean 69 and standard deviation 9, how high does your score have to be to earn an A?

Activity 15-7: Students' Measurements *(cont.)*

Consider the data collected in Topic 3 on the heights and armspans of each member of your class.

(a) Enter these data into the calculator, and use the calculator to compute the *ratio* of height to armspan (i.e., height divided by armspan) for each student. Produce (by hand) a dotplot of the distribution of these ratios.
(b) For what proportion of the students in this sample is height greater than armspan? What can you say about the value of the height/armspan ratio for these students?
(c) Does the distribution of ratios appear to be roughly normal?
(d) Have the calculator compute the mean and standard deviation of these ratios. Report these values.
(e) Suppose that these ratios in the population of all college students do in fact follow a normal distribution with mean and standard deviation equal to those found for this sample. Under this assumption, calculate the proportion of all students who have a ratio greater than one (i.e., height greater than armspan).

Activity 15-8: Candy Bar Weights

Suppose that the wrapper of a candy bar lists its weight as 8 ounces. The actual weights of individual candy bars naturally vary to some extent, however. Suppose that these actual weights vary according to a normal distribution with mean $\mu = 8.3$ ounces and standard deviation $\sigma = 0.125$ ounces.

(a) What proportion of the candy bars weigh less than the advertised 8 ounces?
(b) What proportion of the candy bars weigh more than 8.5 ounces?
(c) What is the weight such that only 1 candy bar in 1000 weighs less than that amount?

(d) If the manufacturer wants to adjust the production process so that only 1 candy bar in 1000 weighs less than the advertised weight, what should the mean of the actual weights be (assuming that the standard deviation of the weights remains .125 ounces)?

(e) If the manufacturer wants to adjust the production process so that the mean remains at 8.3 ounces but only 1 candy bar in 1000 weighs less than the advertised weight, how small does the standard deviation of the weights need to be?

Activity 15-9: Family Lifetimes

An individual has retraced the genealogy of his family and recorded the ages at death for 35 of his predecessors. He calculates the mean of these lifetimes to be 72.5 years and the standard deviation to be 10.4 years.

(a) Assuming that these lifetimes follow a normal distribution with the same mean and standard deviation as the sample, what would be the probability of an individual family member living to age 80 or higher?

(b) Repeat (a) for the probability of dying before age 60.

(c) Repeat (a) for the probability of living between 75 and 85 years.

Consider a dotplot and boxplot of the distribution of the lifetimes of these 35 family members:

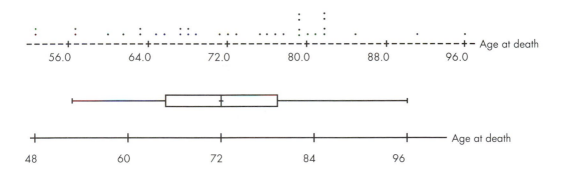

(d) Does this distribution look roughly normal? Do you think the assumption underlying the probability calculations in (a)–(c) is a reasonable one?

Activity 15-10: Cars' Fuel Efficiency *(cont.)*

Suppose that the fuel efficiency (in miles per gallon) of a Honda Civic varies from tankful to tankful according to a normal distribution with mean $\mu = 34$ and standard deviation $\sigma = 3.5$ miles per gallon.

(a) What proportion of all tankfuls would get over 40 miles per gallon?
(b) What proportion of all tankfuls would get less than 30 miles per gallon?
(c) What proportion of all tankfuls would get between 30 and 40 miles per gallon?

Activity 15-11: Climatic Conditions *(cont.)*

In Activity 5-6 you learned that San Diego has an average high temperature in January of 65.9 degrees Fahrenheit. Suppose that the high temperatures on January days in San Diego over the years have followed a normal distribution with mean 65.9. Suppose further that 22% of these high temperatures exceed 75 degrees. Use this information and the table of standard normal probabilities to determine the value of σ, the standard deviation of daily high temperatures in January.

Activity 15-12: SAT's and ACT's *(cont.)*

Refer to the information presented in Activity 5-4 about SAT and ACT tests. Recall that among the college's applicants who take the SAT, scores have a mean of 896 and a standard deviation of 174. Further recall that among the college's applicants who take the ACT, scores have a mean of 20.6 and a standard deviation of 5.2. Consider again applicant Bobby, who scored 1080 on the SAT, and applicant Kathy, who scored 28 on the ACT.

(a) Assuming that SAT scores of the college's applicants are normally distributed, what proportion of applicants score higher than Bobby on the SAT's?
(b) Assuming that ACT scores of the college's applicants are normally distributed, what proportion of applicants score higher than Kathy on the ACT's?

(c) Which applicant seems to be the stronger in terms of standardized test performance? Compare your answer to that of question (g) of Activity 4-4.

Activity 15-13: Empirical Rule

Once again, let Z denote a standard normal distribution. Use the table of standard normal probabilities to find:

(a) $\Pr(-1 < Z < 1)$
(b) $\Pr(-2 < Z < 2)$
(c) $\Pr(-3 < Z < 3)$

This activity demonstrates the theoretical basis of what we called the "empirical rule" in previous topics. With *any* normal distribution, roughly 68% of the observations fall within one standard deviation of the mean, about 95% within two standard deviations of the mean, and about 99.7% within three standard deviations of the mean.

Activity 15-14: Critical Values

Again letting Z denote a standard normal distribution, use the table of standard normal probabilities (in "reverse") to find as accurately as possible the values z^* satisfying:

(a) $\Pr(Z > z^*) = .10$
(b) $\Pr(Z > z^*) = .05$
(c) $\Pr(Z > z^*) = .01$
(d) $\Pr(Z > z^*) = .001$

These values z^* are called the *critical values* of the standard normal distribution.

Activity 15-15: Random Normal Data

The following dotplots display distributions of samples of hypothetical exam scores. Two of these samples were drawn from populations which are normally distributed. Identify the *three* samples which do *not* come from normal populations, and explain in each case why the sample is clearly non-normal. (The data appear in Appendix A.)

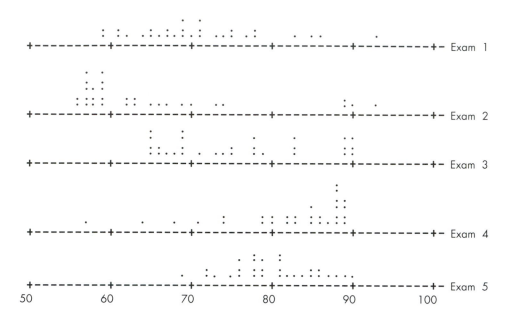

Activity 15-16: Miscellaneous Normal Distributions

(a) Identify at least three variables that you have studied in this text (either data collected in class or supplied in the text) that roughly have a normal distribution.

(b) List at least three variables that you have studied in this text that clearly do not have a normal distribution.

WRAP-UP

This topic has introduced you to the most important mathematical model in all of statistics—the *normal distributions*. You have discovered how to use a *table of standard normal probabilities* and have learned the technique of *standardization*. These skills have enabled you to calculate a variety of probabilities related to normal distributions.

The next topic will ask you to use these skills to determine probabilities related to the sampling distributions of statistics that you studied through simulations in Topics 12 and 13. You will work with the most important theoretical result in all of statistics: the *Central Limit Theorem*.

Topic 16:

CENTRAL LIMIT THEOREM

OVERVIEW

In previous activities you have used actual experiments and calculator simulations to discover that while the value of a sample statistic varies from sample to sample, there is a very precise long-term pattern to that variation. In the last activity you learned how to use normal distributions to perform probability calculations. This topic brings those two ideas together through the *Central Limit Theorem*. This result asserts that the long-term pattern of the variation of a sample proportion is that of a normal distribution. You will examine implications and applications of this theorem in detail, focusing on how it lays the foundation for widely used techniques of statistical inference.

OBJECTIVES

- To understand the *Central Limit Theorem* as describing the (approximate) sampling distribution of a sample proportion.
- To recognize the connection between Central Limit Theorem calculations and the simulations that you performed and analyzed in earlier topics.
- To become proficient with using the normal approximation to calculate probabilities of sample proportions falling in given intervals.
- To discover and appreciate the effect of the sample size on the sampling distribution and on relevant probabilities.
- To gain some insight and experience concerning the use of the Central Limit Theorem for drawing inferences about a population proportion based on a sample proportion.

PRELIMINARIES

1. Take a guess as to the percentage of Americans eligible to vote in the 1992 Presidential election who actually did vote.

2. Of California and Ohio, which state do you suspect has the larger proportion of residents who speak a language other than English at home?

3. Guess the proportion of California residents who speak a language other than English at home.

4. Guess the proportion of Ohio residents who speak a language other than English at home.

5. If 45% of all Reese's Pieces are orange, are you more likely to find less than 40% orange in a sample of 75 candies or in a sample of 175 candies?

6. If 45% of all Reese's Pieces are orange, are you more likely to find between 35% and 55% orange in a sample of 75 candies or in a sample of 175 candies?

IN-CLASS ACTIVITIES

An important theoretical result known as the *Central Limit Theorem* (CLT) confirms what you have already observed about the long-term pattern of variation among sample proportions:

> ## *Central Limit Theorem (CLT) for a Sample Proportion*:
>
> Suppose that a simple random sample of size n is taken from a large population in which the *true* proportion possessing the attribute of interest is θ. Then, provided that n is large (≥ 30 as a rule-of-thumb), the sampling distribution of the *sample* proportion $\hat{p}$ is approximately *normal* with mean θ and standard deviation $\sqrt{\theta(1 - \theta)/n}$.

Together with your knowledge of normal distributions, standardization, and the table of standard normal probabilities, this result enables you to determine the *probability* that a sample proportion will fall within a given interval of values. This is equivalent to finding the proportion of all possible random samples that would produce a sample proportion falling within the interval. This result therefore allows you to answer analytically the types of questions that you studied with simulations in Topics 12 and 13.

Activity 16-1: Sampling Reese's Pieces (*cont.*)

As you did in question (h) of Activity 13-2, consider taking a simple random sample of size $n = 75$ from the population of Reese's Pieces. Continue to assume that 45% of this population is orange ($\theta = .45$).

(a) According to the Central Limit Theorem, how would the sample proportion of orange candies vary from sample to sample? Describe not only the shape of the distribution but also its mean and standard deviation.

The following is a sketch of this distribution; it should resemble very closely the results of your simulated samples from Activity 13-2.

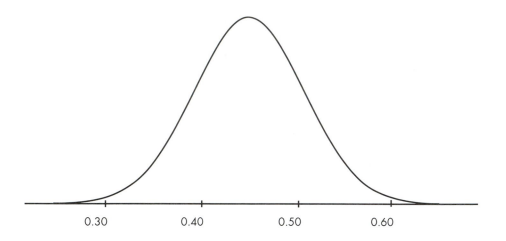

(b) Shade the area under this curve corresponding to the probability that a sample proportion of orange candies will be less than .4. Judging from this area, guess the value of the probability.

(c) Use the technique of standardization and the table of standard normal probabilities to calculate the probability that a sample proportion of orange candies will be less than .4.

The clearest way to record this calculation is as follows:

$$\Pr(\hat{p} < .40) = \Pr\left(Z < \frac{.40 - .45}{.0574}\right) = \Pr(Z < -0.87) = .1922$$

(d) Calculate the probability that a sample proportion of orange candies will fall between .35 and .55 (i.e., within ±.10 of .45). Use the notation introduced in the line above as you find and report the answer.

(e) Compare this probability with question (j) of Activity 12-2, in which you found the percentage of 500 simulated sample proportions which fell within ±.10 of .45. Are they reasonably close?

Activity 16-2: ESP Testing (*cont.*)

Consider the ESP test described in Activity 14-3. Continue to assume that the subject is presented with 30 cards and that she blindly guesses on each card.

(a) If a subject is just guessing, what does the Central Limit Theorem say about the sampling distribution of the sample proportion of correct responses? Report the mean and standard deviation of this distribution as well as describing its shape.

(b) Draw (by hand) a sketch of this sampling distribution on the axis provided below.

```
    0.00        0.15        0.30        0.45        0.60
```

(c) On your sketch above, shade the area corresponding to the probability of a guessing subject getting 38% or more correct responses. Based on this area, guess the value of this probability.

(d) Use the technique of standardization and the table of standard normal probabilities to calculate the probability of a guessing subject getting 38% or more correct responses. Employ the notation illustrated above as you perform the calculations.

(e) Look back at the 10,000 simulated ESP tests presented in Activity 14-3. Compare the probability from (d) with the proportion of the 10,000 simulated tests in which a guessing subject got 38% or more correct responses. Are they reasonably close?

(f) How surprising would it be for a subject to get 38% or more correct responses if he or she were just guessing? Would such an event happen less than 10% of the time in the long run? Would it happen less than 1% of the time in the long run?

Activity 16-3: Effects of Sample Size

Now suppose that you take simple random samples of size $n = 175$ from the population of Reese's Pieces, still assuming that 45% of this population is orange ($\theta = .45$).

(a) According to the Central Limit Theorem, how would the sample proportion of orange candies vary from sample to sample? Describe not only the shape of the distribution but also its mean and standard deviation. What is different from when the sample size was 75?

(b) The following is a sketch of these sampling distributions for sample sizes of 75 and 175; label which is which.

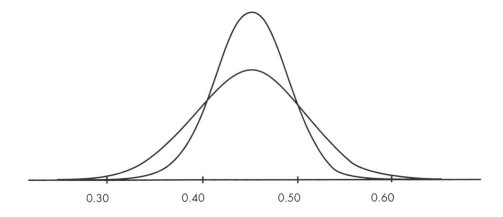

(c) Shade the area corresponding to the probability that a sample proportion of orange candies will be less than .4 (with a sample size of 175). Judging from this area, guess the value of the probability.

(d) Use the technique of standardization and the table of standard normal probabilities to calculate the probability that (with a sample size of 175) a sample proportion of orange candies will be less than .4. How does this probability differ from when the sample size is 75?

(e) Calculate the probability that a sample proportion of orange candies (with a sample size of 175) will fall between .35 and .55 (i.e., within ±.10 of .45). How does this probability differ from when the sample size is 75?

HOMEWORK ACTIVITIES

Activity 16-4: ESP Testing (*cont.*)

Reconsider the ESP test described in Activity 14-3. Suppose now that the subject is presented with $n = 100$ cards.

(a) If the subject is just guessing, will he or she *always* get *exactly* 25 correct? Explain.

(b) According to the Central Limit Theorem, how would the proportion of correct responses vary from subject to subject if all of the subjects were just guessing? Specify the shape of the distribution as well as its mean and standard deviation in presenting your answer. Also sketch the distribution.

(c) Use the CLT and a table of standard normal probabilities to find the probability that a guessing subject would get 27% or more correct out of 100 cards. Also fill in the appropriate area to represent this probability on a sketch of the sampling distribution.

(d) Based on the magnitude of this probability, would you say that it would be very surprising, somewhat surprising, or not terribly surprising for someone to get 27% or more correct even if he or she is blindly guessing on each card?

(e) Repeat (c) and (d) with regard to getting 31% or more correct simply by guessing.

(f) Repeat (c) and (d) with regard to getting 35% or more correct by simply guessing.

(g) Try to calculate how many cards a subject would have to identify correctly in order for the probability of having done that well or better by sheer guessing to be only .025.

Activity 16-5: Voting Behavior

Suppose that you are interested in the proportion of all registered voters who intend to vote in the next election; call this (parameter) proportion θ. In practice, you would never know θ, but you could estimate it based on a sample proportion. Assume for now, though, that you know θ to equal .6, and suppose that you plan to interview a simple random sample of $n = 750$ registered voters and to ask each whether she or he intends to vote.

(a) What does the Central Limit Theorem say about the sampling distribution of this sample proportion? Draw a sketch of this distribution.

(b) Use this result and the table of standard normal probabilities to find the probability that the sample proportion intending to vote would fall within .029 of θ; i.e., between .571 and .629 (still assuming that $\theta = .6$). Also fill in this region on a sketch of the sampling distribution.

(c) Calculate the probability that the sample proportion intending to vote would fall within .035 of θ; i.e., between .565 and .635 (still assuming that $\theta = .6$).

(d) Calculate the probability that the sample proportion intending to vote would fall within .046 of θ; i.e., between .554 and .646 (still assuming that $\theta = .6$).

(e) Try to find the value k such that the probability of the sample proportion falling between $.6 - k$ and $.6 + k$ would equal .8. Fill in the appropriate area on a sketch of the sampling distribution.

Activity 16-6: Presidential Votes *(cont.)*

Recall that 43% of the voters in the 1992 Presidential election voted for Bill Clinton. Suppose that you take a simple random sample of 500 voters from this population.

(a) Is 43% a parameter or a statistic?

(b) Determine the probability that the sample proportion of Clinton voters turns out to be less than 40%.

(c) Determine the probability that the sample proportion of Clinton voters exceeds 50%.

(d) Determine the probability that the sample proportion of Clinton voters falls between 40% and 46%.

(e) Determine the probability that the sample proportion of Clinton voters falls between 37% and 49%.

(f) Determine the probability that the sample proportion of Clinton voters falls between 43% and 89%.

(g) Without doing the actual calculations, indicate how your answers to (b)–(f) would change (get smaller, get larger, or stay the same) if the sample size were 1500 instead of 500.

Activity 16-7: Widget Manufacturing *(cont.)*

Suppose that when a retailer receives a large shipment of widgets from the manufacturer, they inspect a simple random sample of 200 of the items. Suppose further that the retailer accepts the shipment if 10% or fewer of the *sampled* items are defective.

(a) Determine the probability that they will accept the shipment if it is actually the case that 13% of the *population* is defective.
(b) Determine the probability that they will accept the shipment if it is actually the case that 15% of the *population* is defective.
(c) Determine the probability that they will accept the shipment if it is actually the case that 20% of the *population* is defective.
(d) Determine the probability that they will *not* accept the shipment even if it is actually the case that only 8% of the *population* is defective.

Activity 16-8: Boy and Girl Births *(cont.)*

Reconsider Activities 12-8 and 13-10 which asked about births in two local hospitals: hospital A which has 100 deliveries per day and hospital B which has 20 deliveries per day. Continue to assume that each birth is equally likely to be a boy or a girl.

(a) Use the Central Limit Theorem to sketch the sampling distributions of daily sample proportions of boy births in each hospital. Sketch them on the same scale.
(b) Use standardization and the table of standard normal probabilities to determine the probability of hospital A having 60% or more boy births in a day.
(c) Use standardization and the table of standard normal probabilities to determine the probability of hospital B having 60% or more boy births in a day.
(d) Again answer which hospital is more likely to have a day for which 60% or more of its births are boys. Compare your answer now to your answers in Activities 12-8 and 13-10.

Activity 16-9: Sampling Reese's Pieces *(cont.)*

As in Activities 16-1 and 16-3, consider again sampling Reese's Pieces, and continue to assume that the population proportion of orange candies is $\theta = .45$. Determine how large the sample size n must be so that the probability of obtaining 40% or fewer orange candies in the sample is .025. Use the Central Limit Theorem, standardization, and the table of standard normal probabilities to determine this.

Activity 16-10: Presidential Votes *(cont.)*

According to the Voter News Service, 55.9% of those who were eligible actually voted in the 1992 Presidential election. Suppose that you could take simple random samples of adult Americans.

(a) According to the Central Limit Theorem, how would the sample proportion of voters vary from sample to sample? Draw a sketch of this distribution as well as specifying it in words.

(b) Suppose that the simple random sample contains 250 adult Americans. Use the CLT to determine the probability that less than 50% of those in the sample would have voted.

(c) Repeat (b) if the simple random sample consists of 100 adult Americans.

(d) Repeat (b) if the simple random sample consists of 750 adult Americans.

Activity 16-11: Non-English Speakers

In the state of California in 1990, 31.5% of the residents spoke a language other than English at home. Suppose that you take a simple random sample of 100 California residents.

(a) Determine the probability that more than half of the residents sampled would speak a language other than English at home.

(b) Determine the probability that less than one-quarter of the residents sampled would speak a language other than English at home.

(c) Determine the probability that between one-fifth and one-half of the residents sampled would speak a language other than English at home.

(d) Without doing the calculations, indicate how you would expect the answers to (a), (b), and (c) to change for a simple random sample from the state of Ohio, where 5.4% of the 1990 residents spoke a language other than English at home.

(e) Draw sketches of the California and Ohio sampling distributions on the same scale.

Activity 16-12: Home Court Advantage *(cont.)*

Refer to Activity 14-8, where you considered the question of whether the home team wins more than half of its games in the National Basketball Association. Suppose that the home team actually wins 50% of all games and that you study a simple random sample of 30 NBA games.

(a) According to the Central Limit Theorem, how would the *sample* proportion of home victories vary from sample to sample?

(b) Determine the probability that the home team would win 60% or more of the games in a sample of 30 games (still assuming that the home team wins 50% of all games). Is this probability less than .10?

(c) Of the 30 NBA games played on January 3–7, 1992, 18 were won by the home team. In light of your answer to (b), does this sample information provide strong evidence of a home court advantage?

(d) Suppose that the home team actually wins 60% of all NBA games. What then would be the probability that the home team would win 60% or more of the games in a sample of 30 games?

Activity 16-13: Home Court Advantage *(cont.)*

Refer to Activity 16-12. Suppose that the study had followed 120 NBA games and found that the home team won 72 (60%) of them.

(a) Without doing any calculations, how would you expect the probability in (b) of Activity 16-12 to change with this larger sample size? Explain your answer.

(b) Perform the probability calculation asked for in (b) of Activity 16-12 with this larger sample size. Does the calculation confirm your answer to (a)?

(c) How would the conclusion asked for in (c) of Activity 16-12 change with this larger sample size? Explain.

Activity 16-14: Mother's Day Cards

Suppose that 80% of all American college students send a card to their mother on Mother's Day. Suppose further that you plan to select a simple random sample of 300 American college students and to determine the proportion of them who send a card to their mother on Mother's Day.

(a) Is 80% a parameter or a statistic? What symbol have we used to represent it?

(b) According to the Central Limit Theorem, how would the sample proportion who send a card to their mother vary from sample to sample?

(c) Determine the probability that less than three-fourths of the sample send a card to their mother.

(d) Would the answer to (c) be smaller, larger, or the same if a sample size of 800 was used? Explain your answer without performing the calculation.

(e) One can show that in this context $\Pr(p > .85) \approx .015$. Write a sentence or two explaining for a layperson what this statement means.

WRAP-UP

This topic has given you a great deal of practice using the result of the *Central Limit Theorem* to calculate probabilities that a sample proportion will fall in a certain interval (assuming that the population proportion is known). It has also tried to reinforce the fundamental idea that sampling distributions have less variability with larger sample sizes.

In practice, of course, one wishes to go in the other direction: to make inferences about a population parameter based upon the observed value of a sample statistic. Even though these two approaches seem to pull in opposite directions, the Central Limit Theorem actually provides the justification for much of *statistical inference*.

Without specifying them by name, this topic has also explored ideas related to *statistical confidence* and *statistical significance*. These are the two most important concepts in statistical inference; understanding them is crucial to being able to produce and to interpret statistical reasoning. These two ideas form the focus of the next unit of the text.

Unit Four

Inference from Data: Principles

Topic 17:

CONFIDENCE INTERVALS I

In the last unit you explored how sample statistics (in particular, sample proportions) vary from sample to sample. The Central Limit Theorem has allowed you to make probability statements about a sample proportion falling in a certain interval, provided that one knows the value of the population proportion. The much more common problem is to estimate or to make a decision about an unknown population parameter based on an observed sample statistic. These are goals of *statistical inference*.

There are two major techniques of classical statistical inference: *confidence intervals* and *tests of significance*. Confidence intervals seek to estimate a population parameter with an interval of values calculated from an observed sample statistic. Tests of significance assess the extent to which sample data support a particular hypothesis concerning a population parameter. This topic extends your study of the concept of statistical confidence by introducing you to the construction of confidence intervals for estimating a population proportion.

- To understand the purpose of *confidence intervals* for estimating population parameters.
- To learn how to construct confidence intervals for estimating a population proportion.
- To appreciate the distinctions between correct and incorrect interpretations of confidence intervals.

- To explore some properties of confidence intervals related to confidence level and sample size.
- To calculate confidence intervals for a variety of genuine applications and interpret the results.

PRELIMINARIES

1. If a new penny is *spun* on its side (rather than tossed in the air), about how often would you expect it to land "heads" in the long run?

2. Spin a new penny on its side five times. Make sure that the penny spins freely and falls naturally, without hitting anything or falling off the table. How many heads did you get in the five spins?

3. Pool the results of the penny spinning experiment for the entire class; record the total number of spins and the total number of heads.

4. Take a guess as to what proportion of students at your school wear glasses in class.

5. Mark on the number line below an interval which you believe with 80% confidence to contain the actual value of this population parameter (proportion of students at your school who wear glasses in class).

```
+---------+---------+---------+---------+---------+-
0.00      0.20      0.40      0.60      0.80      1.00
```

6. Mark on the number line below an interval which you believe with 99% confidence to contain the actual value of this population parameter.

```
+---------+---------+---------+---------+---------+-
0.00      0.20      0.40      0.60      0.80      1.00
```

7. Which of these two intervals is wider?

8. Record below the number of students in your class and the number who are wearing glasses.

9. Take a guess concerning the proportion of American households that made a financial contribution to a charity in 1989.

10. Think of a popular magazine in which you have an interest. Name this magazine and make a guess concerning the proportion of its pages that contain at least one advertisement.

11. Record below the number of students in your class who have their own credit card and the number who do not.

12. Before the next class meeting, ask at least five *students* from your school whether or not they have their own credit card. Also keep track of their gender.

13. In Activities 17-1, 17-3, and 17-4 you will be experimenting with penny spinning. In Activity 17-4 you are asked to record the results of a calculator simulation of 200 experiments of 80 spins of a penny. In order for your calculator to do this simulation, you need to download SIMINT.83p into your calculator.

    Choose a partner (if you do not have one already) and use one of the calculators to run SIMINT.83p and have the other available for calculations. Enter 80 for the number of spins of a penny, 200 for the number of experiments, 0.95 for the confidence level (which you will learn about during the In-Class Activities), and .35 for θ (theta), the assumed population proportion. The information given on your home screen, while the program is running, tells you the number of experiments that have been completed, the upper and lower bounds of the confidence interval (again, which you will learn about), and whether the true population proportion θ falls within the interval. The simulation will take approximately 12 minutes.

IN-CLASS ACTIVITIES

Activity 17-1: Penny Spinning

(a) Determine the proportion of the class penny spins that landed heads.

(b) Is this proportion a parameter or a statistic? What symbol would you use to represent it?

The parameter in this example, denoted by θ, would be the true proportion of spins that would land "heads" among the (hypothetically infinite) population of all possible spins of a penny.

(c) Does the class experiment allow us to determine θ exactly?

(d) Is θ more likely to be close to the observed proportion of heads than to be far from it?

These questions deal with the issue of *statistical confidence* that you explored in Topic 13. While a sample statistic provides a reasonable estimate of a population parameter, one certainly does not expect the sample statistic to equal the (unknown) population parameter exactly. It is likely, however, that the unknown parameter value is "in the ballpark" of the sample statistic. The purpose of *confidence intervals* is to use the sample statistic to construct an *interval* of values which one can be reasonably *confident* contains the true (unknown) parameter.

There are two parts to every confidence interval. One is the interval itself; the other is the *confidence level*, which is the measure of how confident one is that the interval does in fact contain the true parameter value. One specifies the confidence level by deciding the level of confidence that is necessary for the given situation; common values are 90%, 95%, and 99%.

The interval itself is generated by starting with the sample statistic and moving a certain distance on either side of it. This distance depends on the confidence level desired and is based on the Central Limit Theorem, which tells one how much variability to expect from sample to sample.

In the context of estimating a population proportion θ from a sample proportion $\hat{p}$, it turns out that the confidence interval takes the form

$$\hat{p} \pm z^* \sqrt{\frac{\hat{p}(1-\hat{p})}{n}}$$

where $\hat{p}$ denotes the sample proportion, n is the sample size, and z^* represents the *critical value* from the standard normal distribution for the confidence level desired. (The "$\pm$" means that one subtracts the term following it from the term preceding it to get the lower endpoint of the interval and then adds the term following it to the term preceding it to get the upper endpoint of the interval.)

The technical assumptions without which this procedure is invalid are that the sample is a simple random sample from the population and that the sample size is large ($n \geq 30$ as a rule-of-thumb).

Activity 17-2: Critical Values

Find the value (call it z^*) such that the area under the standard normal curve between $-z^*$ and z^* is equal to .95 by following the steps below:

(a) Draw a sketch of the standard normal curve and shade the area that is being asked for.

(b) Based on your sketch, what is the *total* area under the standard normal curve to the *left* of the value z^*?

(c) Use the table of standard normal probabilities to find the value (z^*) that has this area (your answer to (b)) to its left under the standard normal curve.

This value z^* is called the *upper .025 critical value* of the standard normal distribution because the area to its right under the standard normal curve is .025. This critical value is used for 95% confidence intervals. Critical values for any confidence level can be found in this manner. Some commonly used critical values are listed in the following table (at the top of the next page):

Confidence level	80%	90%	95%	99%	99.90%
Area to right	0.1	0.05	0.025	0.005	0.0005
Critical value z^*	1.282	1.645	1.96	2.576	3.291

(d) Find (by hand, following the same steps as above) the critical value for an 85% confidence interval.

Activity 17-3: Penny Spinning (*cont.*)

(a) For the penny spinning data collected in class, construct a 95% confidence interval for θ, the true proportion of all (hypothetical) penny spins that would land heads.

(b) Can you be *certain* that this interval contains the actual value of θ?

(c) What is the *width* of this interval? (The width is the difference between the upper and lower endpoints of the interval.)

(d) What is the *half-width* of this interval?

(e) Explain how you could have determined the half-width of the interval without first having calculated the width.

(f) What assumptions underlie the validity of the procedure? Are they satisfied here? If not, is the violation serious enough to cause you to doubt the validity of the interval?

Activity 17-4: Calculator Simulations

Assume for the moment that 35% of all spun pennies land heads; i.e., that the population proportion of heads is $\theta = .35$. To examine more closely what the interpretation of "confidence" means in this statistical context, we asked you in the preliminaries section to use the calculator to simulate 200 experiments of 80 spins each, assuming that $\theta = .35$. The calculator constructed a 95% confidence interval for θ for each of the sample proportions of heads obtained.

(a) How many of the 200 confidence intervals actually contain the actual value (.35) of θ? What percentage of the 200 experiments produce an interval which contains .35? (If you would like to view the results of this simulation, the STAT LIST EDITOR has been set up to include the upper and lower bounds of the interval, and the proportions).

(b) For each interval that does not contain the actual value (.35) of θ, record its sample proportion of $\hat{p}$ heads obtained (this information is stored in a list named PROP). What do these values have in common?

(c) If you had conducted a *single* experiment of 80 penny spins without knowing the population proportion θ, would you have any *definitive* way of knowing whether your 95% confidence interval contained the actual value of θ?

(d) Explain the sense in which you would be 95% confident that your 95% confidence interval contains the actual value of θ.

This simulation illustrates the following concerns about the proper use and interpretation of confidence intervals:

- One interprets a 95% (for instance) confidence interval by saying that one is 95% *confident* that the interval contains the true value of the population proportion.
- More specifically, this interpretation means that if one repeatedly takes simple random samples from the population and constructs 95% confidence intervals for each sample, then in the long run 95% of these confidence intervals will contain the true population proportion.
- It is technically incorrect to say that the probability is .95 that a 95% confidence interval contains the true population proportion. The technicality is that the population proportion is not random; it is some fixed (but unknown) value. What *is* random is the sample proportion and thus the *interval* based on it.
- Confidence intervals estimate the value of a *parameter;* they do not estimate the value of a statistic or of an individual observation.

Activity 17-5: Effect of Confidence Level

Consider again the penny spinning experiment, the sample results obtained in class, and the issue of trying to estimate the true value of θ based on the sample results.

(a) *Intuitively* (without thinking about the formula), would you expect a 90% confidence interval for the true value of θ to be wider or narrower than a 95% interval? (Think about whether the need for greater confidence would produce a wider or a narrower interval.) Explain your thinking.

(b) To investigate your conjecture you can use the 1-PropZInt feature of your calculator. Press $\boxed{\text{STATS}}$ and then choose the "TESTS" option. Scroll down to the 1-PropZInt command and press $\boxed{\text{ENTER}}$. For example, if 40 out of 100 spins are heads and you want to produce a 95% confidence interval, then you would enter the following into your calculator:

```
1-PropZInt
 x:40
 n:100
 C-Level:.95
 Calculate
```

Highlight "Calculate" and press $\boxed{\text{ENTER}}$. Your home screen should read as:

```
1-PropZInt
 (.30398,.49602)
 p̂=.4
 n=100
```

Use the calculator to produce 80%, 90%, 95% (which you have already done by hand), and 99% confidence intervals for θ. Record the results in the table and also mark the intervals on the scale below:

Confidence level	Confidence interval	Half-width	Width
80%			
90%			
95%			
99%			

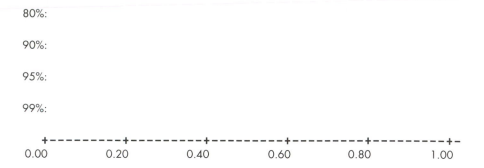

80%:

90%:

95%:

99%:

```
    +---------+---------+---------+---------+---------+-
  0.00      0.20      0.40      0.60      0.80      1.00
```

(c) Do you need to modify your answer to (a) in light of these findings?

Activity 17-6: Effect of Sample Size

Continue to consider the penny spinning experiment, but suppose that you could call for different numbers of pennies to be spun (i.e., different sample sizes).

(a) How would you *intuitively* (again, without considering the formula) expect the width of the 95% confidence interval to be related to the sample size used? For example, would you expect the interval to be wider or narrower if 500 pennies were spun as opposed to 100 pennies? Explain your thinking.

(b) To investigate this question, use the 1-PropZInt feature of your calculator to produce a 95% confidence interval using each of the sample sizes listed in the table. Fill in the table and also mark the intervals on the scale below:

Sample size	Sample heads	Confidence interval	Half-width	Width
100	35			
400	140			
800	280			
1600	560			

n = 100:

n = 400:

n = 800:

n = 1600:

(c) Do you need to modify your answer to (a) in light of these findings?

(d) How does the half-width when *n* = 100 compare to the half-width when *n* = 400? How does the half-width when *n* = 400 compare to that when *n* = 1600?

(e) What effect does *quadrupling* the sample size have on the half-width of a 95% confidence interval?

HOMEWORK ACTIVITIES

Activity 17-7: Students' Glasses in Class

(a) Recall the information collected from students in your class concerning the wearing of glasses in class. What proportion of the class wears glasses? Is this number a parameter or a statistic?

(b) Observe at least 20 additional students and note how many wear glasses during class. (Try not to include anyone from your class again.) Record the number that you observe and the number who wear glasses in class.

(c) Combine the sample results of your class and the other students whom you observe. What is the total number of these subjects and the total number who wear glasses? What proportion of students in this combined sample wear glasses? What symbol is appropriate for representing this proportion?

(d) Use this (combined) sample result to produce a 90% confidence interval for θ, the proportion of *all* students at your school who wear glasses in class.

(e) Write a sentence or two interpreting this interval; be sure to relate the interpretation to the context.

Activity 17-8: Closeness of Baseball Games

The following table tallies the margin of victory in the 355 Major League baseball games played during June of 1992:

Margin	1	2	3	4	5	6	7
Tally	106	64	54	41	26	26	14
Margin	8	9	10	11	12	13	14
Tally	6	6	5	3	0	2	2

(a) What proportion of the games were decided by one run? Is this number a parameter or a statistic?

(b) Let θ denote the true proportion of *all* Major League baseball games that are decided by one run. Is θ a parameter or a statistic?

(c) Use the sample result to estimate θ with a 90% confidence interval, with a 95% confidence interval, and with a 99% confidence interval. Do these by hand, but feel free to use the calculator to check your work.

(d) Write a sentence or two describing what these intervals reveal about the proportion of baseball games decided by a single run.

(e) Is this sample a simple random one from the population of all Major League baseball games? If not, is the sample likely to be biased in a certain direction with respect to margin of victory? Explain.

(f) If the games played in June of 1992 were again used as the sample but game-time temperature were the variable of interest, would the sample likely be biased in a certain direction? Explain.

Activity 17-9: Magazine Advertisements

(a) Find a recent issue of the magazine that you mentioned in the Preliminaries section. Take a simple random sample of 30 pages from the issue by using either the calculator or a table of random digits. Record each page number and whether or not the page contains at least one advertisement.

(b) What proportion of this sample of pages contains at least one advertisement?

(c) Find (by hand) a 95% confidence interval for θ, the *true* proportion of *all* of the magazine pages that contain at least one advertisement.

(d) Does the interval contain your guess from the "Preliminaries" section?

(e) Does the interval contain the value .5?

Activity 17-10: Television Magic

In June 1994, the cable television show *Nick-at-Nite* conducted a phone-in survey of its viewers with regard to the question: "Which classic sitcom character—Jeannie from *I Dream of Jeannie* or Samantha from *Bewitched*—possessed more magic powers?" Samantha received 810,000 votes to Jeannie's 614,000.

(a) Identify the population and the parameter of interest in this case.

(b) Is this a simple random sample from the population of interest?

(c) Find (by hand) a 97% confidence interval for the population parameter of interest.

(d) Explain why this confidence interval is so narrow.

Activity 17-11: Cat Households *(cont.)*

A survey of 40,000 American households in 1987 found that 30.5% of the households in the sample owned a pet cat.

(a) Use this sample information to form a 99% confidence interval to estimate the proportion of all American households that owned a pet cat in 1987.
(b) Write a sentence or two interpreting what this interval means.
(c) What additional information about this survey would you need in order to comment on whether the technical assumptions underlying the confidence interval procedure are satisfied in this case?

Activity 17-12: Charitable Contributions

In a survey of American households, 75.1% of the households claimed to have made a financial contribution to charity in the past year.

(a) Describe the parameter value of interest in this situation.
(b) If the survey had involved 1000 households, what would a 99% confidence interval for this parameter be?
(c) If the survey had involved 5000 households, what would a 99% confidence interval for this parameter be?
(d) If the survey had involved 10,000 households, what would a 99% confidence interval for this parameter be?
(e) If the survey had involved 100,000 households, what would a 99% confidence interval for this parameter be?

Activity 17-13: Marriage Ages *(cont.)*

Reconsider Activity 3-7, in which you analyzed the ages of couples who applied for marriage licenses. Consider now the issue of estimating the proportion of all marriages in this county for which the bride is younger than the groom. These 24 couples are actually a subsample from a larger sample of 100 couples who applied for marriage licenses in Cumberland County, Pennsylvania in 1993. Marriage ages for this larger sample appear below:

Husband	Wife	Husband	Wife	Husband	Wife	Husband	Wife	Husband	Wife
22	21	40	46	23	22	31	33	24	25
38	42	26	25	51	47	23	21	25	24
31	35	29	27	38	33	25	25	46	37
42	24	32	39	30	27	27	25	24	23
23	21	36	35	36	27	24	24	18	20
55	53	68	52	50	55	62	60	26	27
24	23	19	16	24	21	35	22	25	22
41	40	52	39	27	34	26	27	29	24
26	24	24	22	22	20	24	23	34	39
24	23	22	23	29	28	37	36	26	18
19	19	29	30	36	34	22	20	51	50
42	38	54	44	22	26	24	27	21	20
34	32	35	36	32	32	27	21	23	23
31	36	22	21	51	39	23	22	26	24
45	38	44	44	28	24	31	30	20	22
33	27	33	37	66	53	32	37	25	32
54	47	21	20	20	21	23	21	32	31
20	18	31	23	29	26	41	34	48	43
43	39	21	22	25	20	71	73	54	47
24	23	35	42	54	51	26	33	60	45

(a) For how many of these 100 marriages can you determine which partner was younger? (In other words, eliminate the cases in which both bride and groom listed the same age on the marriage license since you cannot tell which partner is younger in those cases.)

(b) In how many of *these* marriages is the bride younger than the groom?

(c) Find a 90% confidence interval for the proportion of all marriages in this county for which the bride is younger than the groom.

(d) Find a 95% confidence interval for the proportion of all marriages in this county for which the bride is younger than the groom.

(e) Find a 99% confidence interval for the proportion of all marriages in this county for which the bride is younger than the groom.

(f) Do any of these intervals include the value .5?

(g) Comment on whether the sample data suggest that the bride is younger than the groom in more than half of the marriages in this county.

Activity 17-14: Age and Political Ideology *(cont.)*

Refer to Activity 11-2, in which you studied the relationship between age and political ideology in a sample of 1442 American adults, 704 of whom identified themselves as political moderates.

(a) Produce (either by hand or with the calculator) a 90% confidence interval for the proportion of all American adults who identify themselves as political moderates.
(b) Does this interval include the value .5?
(c) Comment on whether this sample provides any reason to doubt the proposition that half of all American adults consider themselves political moderates.

Activity 17-15: Students' Travels *(cont.)*

Consider again the data from Topic 2 on whether or not students have been to Europe.

(a) Find a 90% confidence interval to estimate the proportion of all students at the college who have been to Europe. Also identify the interval's half-width.
(b) If the sample had been four times as large as it actually was (and the same proportion of the sample had been to Europe), what would the half-width of a 90% confidence interval be? Answer this *without* performing the calculation, and explain how you arrive at the answer.
(c) If the sample had been 16 times its actual size (and the same proportion of the sample had been to Europe), what would the half-width of a 90% confidence interval be? Again answer this *without* performing the calculation, and explain how you arrive at the answer.

WRAP-UP

This topic has provided your first exposure to *statistical inference* by introducing you to *confidence intervals*, a very widely used inference technique. In addition to discovering how to construct confidence intervals for a population proportion, you have seen proper and improper interpretations of them and explored the effect of the confidence level and the sample size on the interval.

In the next topic you will continue to study confidence intervals as they relate to the *margin-of-error* of a sample survey. In following topics you will explore the other major type of inference procedure—*tests of significance*.

Topic 18:

CONFIDENCE INTERVALS II

OVERVIEW

In the previous topic you began a study of the important and widely used technique of *confidence intervals*. These procedures use a sample statistic to estimate a population parameter with an interval of values and a certain confidence level. This topic asks you to continue this study by considering sample survey results and the connection between confidence intervals and the commonly used expression *"margin of error."* You will also investigate some of the limitations of confidence intervals, encountering some situations in which one must be wary of applying them thoughtlessly.

OBJECTIVES

- To understand the use of the term *"margin-of-error"* as it relates to sample surveys.
- To discover how to choose a sample size in order to achieve a confidence interval of a prespecified width at a certain confidence level.
- To appreciate the effect (or lack thereof) of the size of the population on confidence intervals.
- To learn to recognize situations in which confidence intervals cannot be applied meaningfully.
- To recognize the importance of random sampling in the context of confidence intervals.

PRELIMINARIES

1. Tally below the responses of students to the question of whether or not they have their own credit card. Tally separately for men and women.

Men, yes	Men, no	Women, yes	Women, no

2. Count the sample *total* who do and who do not have their own credit card.

3. If a polling organization wants to estimate the proportion of all adult Americans who agree with a certain proposition to within ±0.04 with 95% confidence, guess how many people they would have to sample.

4. If this same polling agency needed to be more confident that the sample proportion would fall within ±0.04 of the population proportion, would they need to sample more or fewer people than in question 3?

5. If this same polling agency wants to estimate the population proportion to within ±0.02 with 95% confidence, would they need to sample more or fewer people than in question 3?

6. Guess what proportion of adult Americans would answer in the affirmative to the question of whether "most U.S. representatives deserve to be re-elected."

7. Think about your local telephone book and take a guess as to the proportion of the individual listings that contain women's names.

IN-CLASS ACTIVITIES

Recall from the previous topic that a confidence interval for a population proportion θ is formed from a sample proportion $\hat{p}$ through the expression

$$\hat{p} \pm z^* \sqrt{\frac{\hat{p}(1 - \hat{p})}{n}}$$

where n is the sample size and z^* is the critical value from the standard normal distribution for the confidence level desired.

Activity 18-1: American Moral Decline (*cont.*)

Recall from Activity 13-7 that a survey conducted by *Newsweek* on June 2–3, 1994 asked adult Americans, "Do you think the United States is in a moral and spiritual decline?" Of the 748 adults surveyed, 76% answered "yes."

(a) Use this sample result to produce (by hand) a 95% confidence interval for θ, the true proportion of *all* adult Americans who would have answered "yes" to the question about moral decline.

(b) What are the formal assumptions which underlie the validity of this procedure?

(c) What is the half-width of this confidence interval?

(d) The *Newsweek* article states that the survey's margin of error is ±3 percentage points. Can you figure out where this number comes from?

The *margin-of-error* of a sample survey refers to (usually, at least) the half-width of a 95% confidence interval for the population proportion of interest.

(e) Remembering the comments about interpreting confidence intervals listed in Topic 17, identify each of the following statements as true or false:

• The survey's margin-of-error *guarantees* that the population proportion θ is within ±.03 of .76.

• The probability is .95 that θ falls within the confidence interval.

• If one repeatedly took random samples of 748 adult Americans, asked them the survey question, and formed confidence intervals in this same manner, then in the long run 95% of the intervals so generated would contain the true value of the population proportion θ.

• If one repeatedly took random samples of 748 adult Americans, asked them the survey question, and formed confidence intervals in this same manner, then in the long run 95% of the intervals so generated would contain the value .76.

(f) Would you expect a survey's margin of error to increase or to decrease as the sample size increases?

(g) Evaluate your conjecture by determining (by hand) the margin-of-error for the survey above if 2000 adult Americans had been interviewed (and assuming that 76% was still the sample proportion of "yes" responses). Then repeat for a sample size of 400.

(h) Do you need to modify your conjecture based on these findings?

(i) If you want to use this survey's results to estimate the proportion of all adult American *men* who would have answered "yes" to the question about moral decline, would the margin-of-error for this proportion be greater or less than ±3%? Explain your answer.

(j) If you want to use this survey's results to estimate the proportion of all adult American male *college graduates* who would have answered "yes" to the question about moral decline, would the margin-of-error for this proportion be greater or less than that for the previous question? Again explain your answer.

These last questions reveal that a survey's margin-of-error increases when one considers subgroups of the population.

Activity 18-2: Congressional Term Limits

Suppose that you are interested in estimating the proportion of American adults who favor the imposition of term limit restrictions on members of Congress (call this proportion θ). Suppose that you want to be able to estimate θ to within .04 with 95% confidence. With this goal in mind, you can determine, *prior to* doing the interviewing, how many people you need to interview to achieve the desired levels of accuracy and confidence.

(a) Using an initial guess that the sample proportion favoring term limits will be in the neighborhood of .5, use the confidence interval expression to determine how many people you would need to sample in order to estimate θ to within ±.04 with 95% confidence. (Be very careful to avoid rounding off numbers in intermediate calculations; such round-off errors can affect further calculations considerably. Also be sure to express your answer as an *integer* number of people.)

(b) Would you expect to need more or fewer people if you wanted to be able to estimate θ to within ±.01 with 95% confidence?

(c) Use the confidence interval expression to determine the sample size required in (b).

(d) Would you expect to need more or fewer people if you wanted to be able to estimate θ to within ±.01 with 99% confidence?

(e) Use the confidence interval expression to determine the sample size required in (d).

(f) Suppose that the population of interest is all adult *Pennsylvanians* instead of all adult Americans. How does this change in the size of the *population* affect your calculations in (a), (c), and (e)?

This last question points out the very surprising (to most people) result that, as long as one is dealing with populations much larger than the samples involved, the margin-of-error of a survey does not depend on the *population* size.

(g) You have seen that certain relationships exist among the sample size of a study, the confidence level desired, and the half-width of the confidence interval. Fill in the following table to summarize how these components interact. For instance, Case 1 asks you to indicate what happens to the confidence interval's half-width as the confidence level increases for a fixed sample size.

Case	Sample size	Confidence level	C.I. half-width
1	Fixed	Increases	
2	Fixed		Increases
3	Increases	Fixed	
4		Fixed	Increases
5	Increases		Fixed
6		Increases	Fixed

(h) How many people would you have to interview in order to determine θ exactly (with 100% confidence)?

Activity 18-3: Female Senators (*cont.*)

Suppose that an alien lands on Earth, notices that there are two different sexes of the human species, and wants to estimate the proportion of all humans who are female. If this alien were to use the members of the 1994 United States Senate as a sample from the population of human beings, it would have a sample of 7 women and 93 men.

(a) Use this sample information to form (either by hand or using the calculator) a 95% confidence interval for the actual proportion of all humans who are female.

(b) Is this confidence interval a reasonable estimate of the actual proportion of all humans who are female?

(c) Explain why the confidence interval procedure fails to produce an accurate estimate of the population parameter in this situation.

(d) It clearly does not make sense to use the confidence interval in (a) to estimate the proportion of women in the world, but does the interval make sense for estimating the proportion of women in the U.S. Senate in 1994? Explain your answer.

This example illustrates some important limitations of the widely used technique of confidence intervals. First, confidence intervals do not compensate for the problems of a biased sample. If the sample is selected from a population in a biased manner, the ensuing confidence interval will be a biased estimate of the population parameter of interest.

A second important point to remember is that confidence intervals use *sample* statistics to estimate *population* parameters. If the data at hand constitute the entire population of interest, then constructing a confidence interval from the data is meaningless.

United States Senators are clearly not a random sample of the world population. Moreover, these 100 Senators cannot be considered a sample of Senators because they comprise the entire *population* of Senators. One knows precisely that the proportion of women in the population of 1994 U.S. Senators is .07, so it is senseless to construct a confidence interval from these data.

Activity 18-4: College Students' Credit

Reconsider the sample data collected in the "Preliminaries" section concerning college students and credit cards. Ignore, for now, the gender variable.

(a) Calculate the sample proportion of college students who have their own credit card.

(b) Use this sample proportion to construct (either by hand or using the calculator) an 85% confidence interval for the population proportion who have their own credit card.

(c) Comment on the population to which you think this confidence interval pertains. For example, do these sample results generalize to the population of all college students, the population of all college students who study statistics, the population of all students at *your* college, or some other population?

This activity indicates the importance of asking to which population the sample results may be generalized meaningfully. Even though the students in your class do not constitute a random sample from the population of students at your school, it seems unlikely that students studying statistics differ systematically with regard to credit card ownership from other students at the college. Students at your school probably do differ in some important ways from students at other types of schools, however. It would seem unwise to generalize these sample results to the entire population of all college students.

HOMEWORK ACTIVITIES

Activity 18-5: Voter Dissatisfaction with Congress

The October 27, 1994 issue of *USA Today* reports that a nationwide telephone poll of 1007 adults asked whether "most U.S. representatives deserve to be re-elected." The sample proportion responding in the affirmative was 43%.

(a) Determine (by hand) the survey's margin-of-error.
(b) Explain in a sentence or two precisely what this margin-of-error means in the context of this survey.
(c) If you wanted to use the results of this survey to estimate the proportion of white males who agree with the proposition, would the margin-of-error be the same? Explain.

Activity 18-6: American Moral Decline *(cont.)*

Reconsider Activity 18-1 and the *Newsweek* survey about moral decline in America.

(a) If the editors of *Newsweek* wanted to estimate the proportion of all adult Americans who would answer "yes" to the question about moral decline to within ±2% with 95% confidence, how many people would they need to interview? (Use the result of their initial survey to provide a guess for the sample proportion.)
(b) If they wanted to estimate the population proportion to within ±1% with 95% confidence, how many people would they need to interview?
(c) If they wanted to estimate the population proportion to within ±0.5% with 95% confidence, how many people would they need to interview?

(d) These questions reveal that in order to cut a survey's margin-of-error in half, the sample size has to be increased by what factor?

Activity 18-7: Telephone Books

Reconsider the poor alien in Activity 18-3 who got an awful estimate of the proportion of females in the world by using the U.S. Senate as its sample. Suppose that the alien, having learned to distinguish male and female names, decides to use a page from a telephone book as its sample.

(a) Select one page from the white pages of a telephone book. Disregard all listings of businesses, listings which provide only initials, and listings with first names that are not gender-specific (like Pat or Fran). For the listings that you can reasonably identify as male or female, count how many are male and how many are female. Record these values. Also identify the telephone book and page number.

(b) What is the sample proportion of females in your sample?

(c) Use your sample data to form a 95% confidence interval for the actual proportion of all humans who are female.

(d) Does this confidence interval provide a reasonable estimate of the actual proportion of all humans who are female? If not, explain why the confidence interval procedure fails in this situation.

(e) Suggest a practical method for the alien to choose a sample which is likely to produce a reasonable estimate of the proportion of women among the human population.

Activity 18-8: Tax Return Errors

(a) Let θ represent the proportion of all personal income tax returns filed for a certain year that contained arithmetic errors. Suppose that the Internal Revenue Service wants to estimate θ to within $\pm.02$ with 95% confidence. Assuming that their initial guess is that θ is in the neighborhood of 30%, how many returns would they need to sample to achieve this goal?

(b) How many returns would they need to sample if they were willing to settle for estimating θ to within $\pm.02$ with 90% confidence?

(c) How many returns would they need to sample if they were willing to settle for estimating θ to within .04 with 90% confidence?

(d) How would your answers to (a)–(c) change if the IRS wanted to estimate the proportion of errors on returns filed by married couples
(rather than the proportion of errors on all returns)?

Activity 18-9: Penny Spinning *(cont.)*

Reconsider the penny spinning experiment of Activity 17-1. Let θ denote
the proportion of *all* spins that would land "heads."

(a) Using the sample proportion obtained in class as an initial estimate, determine how many spins would be required to estimate θ to within $\pm.02$
with 97% confidence.
(b) Determine how many spins would be required to estimate θ to within
$\pm.02$ with 99.9% confidence.
(c) Determine how many spins would be required to estimate θ to within
$\pm.001$ with 99.9% confidence.

Activity 18-10: Emotional Support *(cont.)*

Recall from Activity 12-5 that Shere Hite received mail-in questionnaires
from 4500 women, 96% of whom claimed that they give more emotional
support than they receive from their husbands or boyfriends. Also recall
that an ABC News/*Washington Post* poll of a random sample of 767 women,
found 44% who claimed to give more emotional support than they receive.

(a) Determine the margin-of-error for each of these surveys. Also report
each survey's 95% confidence interval for the proportion of all
American women who feel that they give more emotional support than
they receive.
(b) Are these two confidence intervals similar? Do they overlap at all?
(c) Which survey has the smaller margin-of-error; i.e., the narrower confidence interval?
(d) Which of these two confidence intervals do you have more confidence
in? Explain.
(e) Using the ABC News/*Washington Post* sample proportion as an estimate,
determine how many women you would have to interview in order to
estimate the population parameter to within $\pm.02$ with 98% confidence.

Activity 18-11: Elvis and Alf *(cont.)*

Recall from Activity 12-1 that in 1936 the *Literary Digest* received 2.4 million responses to their survey, with 57% of the sample indicating support for Alf Landon over Franklin Roosevelt in the upcoming election.

(a) Use this sample result to form a 99.9% confidence interval for θ, the actual proportion of all adult Americans who preferred Landon over Roosevelt.
(b) Explain why this confidence interval did such a horrible job of predicting the election result.

Activity 18-12: Veterans' Marital Problems

Researchers recently found that in a sample of 2101 Vietnam veterans, 777 had been divorced at least once.

(a) Use this sample information to form a confidence interval for the proportion of divorced men among *all* Vietnam veterans.
(b) Would it be legitimate to use this interval as an estimate of the divorce rate among all middle-aged American men? Explain.

Activity 18-13: Jury Representativeness

In the legal case of *Swain v. Alabama* (1965), attorneys for Swain, a black man convicted of rape, examined grand jury records in Alabama for the preceding 10 years. Of the 1050 grand jurors over the preceding ten years, 177 had been black. Suppose that the collection of grand jurors could be regarded as a simple random sample from the population of Alabama adults.

(a) With this assumption, find a 99% confidence interval for the proportion of blacks among all Alabama adults.

Blacks actually constituted about 25% of the adult population of Alabama in that time period.

(b) Does the confidence interval in (a) contain the value .25?
(c) Comment on whether the sample data seem to indicate that blacks were

systematically underrepresented on Alabama grand juries in this time period.

Activity 18-14: Penny Thoughts *(cont.)*

Consider again the data collected in Topic 1 concerning whether students would prefer to retain or abolish the penny as a coin of U.S. currency.

(a) What proportion of the students sampled favor retaining the penny?

(b) Use the sample data to find a 90% confidence interval for the proportion of all college students who favor retaining the penny.

(c) What proportion of the students sampled favor abolishing the penny?

(d) Use the sample data to find a 90% confidence interval for the proportion of all college students who favor abolishing the penny.

(e) Is there a connection between the confidence intervals in (b) and in (d)? Explain.

(f) If you wanted to estimate the proportion of all college students who favor retaining the penny with a 99% confidence interval that would have the same half-width as your interval in (a), how many students would you need to sample?

Activity 18-15: Confidence Interval of Personal Interest

Think of a real situation in which you would be interested in producing a confidence interval to estimate a population *proportion*. Describe precisely the population and parameter involved. Also describe how you might select a sample from the population. (Be sure to think of a binary categorical variable so that a proportion is a sensible parameter to deal with.)

Activity 18-16: Confidence Interval of Personal Interest *(cont.)*

Find a newspaper, magazine, or television account of a survey which reports a margin-of-error for the survey. Describe the population involved, the parameter being estimated, and the sampling method used.

WRAP-UP

This topic has extended your study of statistical inference in general and of confidence intervals in particular by examining the term *"margin-of-error"* as it applies to the results of sample surveys.

You have examined both the calculation of margin-of-error and the meaning of the term. You have also discovered how to determine the sample size needed to achieve a desired margin-of-error for a survey. Finally, you have explored situations in which confidence intervals should be applied only with caution.

In the next topic you will continue to study the meaning of *statistical significance* by investigating the other major technique of statistical inference—*tests of significance*.

Topic 19:

TESTS OF SIGNIFICANCE I

The last two topics have introduced you to *statistical inference*, where the goal is to make a statement about a population parameter based on a sample statistic. The *confidence intervals* that you have studied allow one to estimate a population parameter at a certain confidence level with an interval of values.

In this topic you will discover the other major type of statistical inference: *tests of significance*. This procedure assesses the degree to which sample data support a particular conjecture about the value of the population parameter of interest. By exploring the concept of *statistical significance* in an earlier topic, you have already studied the reasoning procedure behind these tests. This topic will introduce you to the formal structure of tests of significance.

- To develop an intuitive understanding of the reasoning process used in *tests of significance*.
- To become familiar with the formal structure of tests of significance and to learn to translate appropriate questions into that structure.
- To discover how to perform calculations relevant to a test of significance concerning a population proportion.

- To acquire the abilities to interpret and to explain the results of tests of significance.
- To explore how the statistical significance of a sample result is related to the sample size of the study.

PRELIMINARIES

1. Recall the ESP test in which a subject has to identify which of four shapes appears on a card. In a test consisting of 100 cards, would you be fairly convinced that a subject Fred does better than just guessing if he gets 28 correct?

2. In the same situation as question 1, would you be fairly convinced that a subject Wilma does better than just guessing if she gets 37 correct?

3. How many of the 100 cards would a subject have to identify correctly before you would be strongly convinced that he or she would do better than just guessing in the long run?

4. If a subject identifies 35% of a sample of cards correctly, would you be more impressed if it were a sample of 200 cards or a sample of 20 cards?

IN-CLASS ACTIVITIES

Activity 19-1: ESP Testing (*cont.*)

Consider again the ESP testing experiment described in Activity 14-3 which you also studied in Activities 16-2 and 16-4. Let θ denote the proportion of correct responses that a particular subject would make in the long run.

(a) If subject Fred is just guessing on each card, what proportion would he get right in the long run? In other words, what value does θ equal for a guesser?

(b) If Fred takes a test consisting of $n = 100$ cards (as in Activity 16-4) and guesses on each card, will he necessarily get exactly 25 correct?

(c) If Fred identifies 28 cards correctly, is it possible that he was just guessing and got a little lucky?

(d) If subject Wilma identifies 37 cards correctly, is it possible that she was just guessing and got lucky?

(e) With which of these two subjects (Fred or Wilma) would you be *more* convinced that the subject does better than just guessing (i.e., would get more than 25% correct in the long run)?

Recall that the concept of *statistical significance* addresses the question of how often a sample result would occur by random variation alone. You studied this idea in Topic 14 through the use of simulations. Now armed with the knowledge of the Central Limit Theorem from Topic 16, you can quantify this idea and calculate the probability that a sample proportion would be as large as the one actually obtained if the subject were just guessing.

(f) Suppose for now that Fred is just guessing (i.e., that $\theta = .25$ for Fred). If one were to repeatedly give him tests of $n = 100$ cards, how does the Central Limit Theorem say that the sample proportions of his correct answers would vary? Indicate the shape of this distribution as well as its mean and standard deviation.

(g) One can use the CLT to determine that the probability is about .2442 of Fred getting 28 or more correct if he is just guessing. Based on this probability, would it be *very surprising* to get a sample proportion of correct answers as high as .28 if the population proportion is $\theta = .25$?

(h) Now suppose that Wilma is just guessing (i.e., that $\theta = .25$ for Wilma) also. One can use the CLT to determine that the probability is about .0028 of her getting 37 or more correct if she is just guessing. Based on this probability, would it be *very surprising* to get a sample proportion of correct answers as high as .37 if the population proportion is $\theta = .25$?

(i) The following table presents four possible sample outcomes. Fill in the missing sample proportions and indicate in the last column how convinced you would be that the subject does better than just guessing (i.e., would get more than 25% correct in the long run). For example, you might use phrases like "not convinced" or "somewhat convinced" or "strongly convinced" to describe your degree of belief.

Subject	Sample number of correct IDs	Sample proportion of correct IDs	Approx. probability of doing so well by just guessing	Your belief that theta > .25 (better than guessing)
Fred	28		.2442	
Barney	31		.0829	
Betty	34		.0188	
Wilma	37		.0028	

This table reveals that your belief in a subject's extrasensory ability should increase as he/she identifies more cards correctly. More important, your degree of belief that $\theta > .25$ (that the subject does better than just guessing) also relates directly to how often the sample result would occur by chance alone if $\theta = .25$. The *smaller* this probability, the *stronger* your belief that $\theta > .25$. In other words, the *smaller* this probability, the stronger the evidence *against* the assumption that $\theta = .25$.

This reasoning is incorporated into a formal structure that one follows when conducting statistical *tests of significance*. This structure involves four essential components and two optional components that are described below:

1. The *null hypothesis* is denoted by H_0. It states that the parameter of interest is equal to a specific, hypothesized value. In the context of a population proportion, H_0 has the form

$$H_0 : \theta = \theta_0,$$

   where θ_0 stands for the hypothesized value of interest. The null hypothesis is typically a statement of "no effect" or "no difference." The significance test is designed to assess the evidence *against* the null hypothesis.

2. The *alternative hypothesis* is denoted by Ha. It states what the researchers suspect or hope to be true about the parameter of interest. It depends on the purpose of the study and must be specified *before* seeing the data. The alternative hypothesis can take one of three forms:

   (a) $H_a: \theta < \theta_0,$ (b) $H_a: \theta > \theta_0,$ or   (c) $H_a: \theta \neq \theta_0$

   The first two forms are called *one-sided* alternatives while the last is a *two-sided* alternative.

3. The *test statistic* is a value computed by *standardizing* the observed sample statistic on the basis of the hypothesized parameter value. It is used to assess the evidence against the null hypothesis. In this context of a population proportion, it is denoted by z and calculated as follows:

$$z = \frac{\hat{p} - \theta_0}{\sqrt{\dfrac{\theta_0(1 - \theta_0)}{n}}}$$

4. The *p-value* is the probability, assuming the null hypothesis to be true, of obtaining a test statistic as extreme or more extreme than the one actually observed. "Extreme" means "in the direction of the alternative hypothesis," so the p-value takes one of three forms (corresponding to the appropriate form of H_a):

   (a) $\Pr(Z < z),$ (b) $\Pr(Z > z),$ or   (c) $2\Pr(Z > |z|)$

   where $|z|$ denotes the absolute value of the test statistic z and Z continues to denote the standard normal distribution.

One judges the strength of the evidence that the data provide against the null hypothesis by examining the p-value. The *smaller* the p-value, the stronger the evidence against H_0 (and thus the stronger the evidence in favor of H_a). For instance, typical evaluations are:

p-value $> .1$: *little* or *no* evidence against H_0
$.05 <$ p-value $\leq .10$: *some* evidence against H_0
$.01 <$ p-value $\leq .05$: *moderate* evidence against H_0
$.001 <$ p-value $\leq .01$: *strong* evidence against H_0
p-value $\leq .001$: *very strong* evidence against H_0

The probability entries in the table above are all p-values. Note that the p-value is not at all the same as "p-hat," the sample proportion $\hat{p}$.

5. (optional) The *significance level*, denoted by α, is an optional "cut-off" level for the p-value that one decides to regard as decisive. The experimenter specifies the significance level in advance; common values are $\alpha = .10$, $\alpha = .05$, and $\alpha = .01$. The smaller the significance level, the more evidence you require in order to be convinced that H_0 is not true.

6. (optional) If the p-value of the test is less than or equal to the significance level α, the *test decision* is to *reject H_0*; otherwise, the decision is to *fail to reject H_0*. Notice that failing to reject is not the same as affirming its truth; it is simply to admit that the evidence was not convincing enough to reject it. Another very common expression is to say that the data are *statistically significant* at the α level if the p-value is less than or equal to α. Thus, a result is statistically significant if it is unlikely to have occurred by chance.

The formal assumptions needed to establish the validity of this significance testing procedure are the same as for the confidence interval procedure: that the data are a simple random sample from the population of interest; and that the sample size is large ($n \geq 30$ as a rule-of-thumb).

Activity 19-2: ESP Testing (*cont.*)

Consider again the ESP testing situation of Activity 19-1. Recall that the researchers seek to establish whether a subject would get more than 25% correct in the long run.

(a) State the *null* hypothesis (in symbols and in words) for the appropriate test of significance in this context.

(b) State the *alternative* hypothesis (in symbols and in words) for the appropriate test of significance in this context.

(c) Recall from the table above that Barney correctly identified the shape on 31 of the 100 cards. Calculate the *test statistic* from this sample information.

(d) Use the table of standard normal probabilities to calculate the p-value of this test. Does your answer agree closely with the probability listed in the table above?

(e) Write a sentence describing what the p-value represents in the context of this experiment.

(f) Based on the p-value, write a one-sentence conclusion regarding the extent to which the data support the claim that Barney does better than just guessing; i.e., that $\theta > .25$ for Barney.

(g) Would you *reject* the *null* hypothesis at the .10 level? at the .05 level? at the .01 level?

(h) Is the sample result *statistically significant* at the .08 level? at the .03 level? at the .005 level?

(i) Fill in the following table to indicate at which significance levels the sample results provide evidence that the subject does better than just guessing.

Subject	Sample proportion	Test statistic	p-value	Signif. at .10 level?	Signif. at .05?	Signif. at .01?
Fred	0.28	0.69	0.2442			
Barney	0.31	1.39	0.0829			
Betty	0.34	2.08	0.0188			
Wilma	0.37	2.77	0.0028			

Activity 19-3: Effect of Sample Size

Suppose that you seek to establish evidence that more than half of all adult Americans have watched a certain television program. Let θ represent the actual proportion of all adult Americans who have watched the program.

(a) Is θ a parameter or a statistic? Explain.

(b) Express (in symbols) the null and alternative hypotheses for addressing the question of interest here.

(c) Suppose that you manage to take a simple random sample of adult Americans and find that 54% of the sample have watched the program. Is 54% a parameter or a statistic?

(d) What further information about the simple random sample would you need to know in order to conduct the appropriate test of significance?

(e) As the sample size increases, would you expect such a sample result ($\hat{p} =$ .54) to provide *stronger* or *weaker* evidence that more than half of all American adults watch the program?

(f) To investigate your conjecture you can use the 1-PropZTest feature of your calculator (located in the STAT TESTS menu). For example, if 51 out of 100 adult Americans have watched a certain television program and you want to find the p-value of the significance test of the null hypothesis that $\theta = .5$ vs. the alternative that $\theta > .5$, then you would set up the 1-PropZTest window of your calculator as:

```
1-PropZTest
 P0:.5
 x:51
 n:100
 Prop≠P0 <P0 >P0
 Calculate Draw
```

Highlight "Calculate" and press ENTER. (Note that the TI-83 uses p_0 rather than θ_0). Your home screen should report the test statistic to be $z = .2$ and the p-value to be .4207 as follows:

```
1-PropZTest
  Prop>.5
  z=.2
  P=.4207403122
  P̂=.51
  n=100
```

Use the calculator to find the p-value of the significance test of the null hypothesis that $\theta = .5$ vs. the alternative that $\theta > .5$. Do this for sample sizes of 100, 300, 500, 1,000, and 2,000. In each case suppose that 54% of the adults in the *sample* watch the program. Record the p-values in the table below and indicate whether the sample result would be considered statistically significant at the significance levels listed. Then indicate whether you need to modify your answer to (e) in light of these results.

Sample size	(One-sided) p-value	Signif. at .10 level?	Signif. at .05 level?	Signif. at .01 level?	Signif. at .001 level?
100					
300					
500					
1000					
2000					

(g) If you read about a survey which reports that 54% of a random sample favors a certain proposition, under what circumstances should you regard this information as strong evidence that more than half of the population favors the proposition?

HOMEWORK ACTIVITIES

Activity 19-4: Home Court Advantage (*cont.*)

Reconsider Activities 14-8 and 16-12, in which you investigated the notion of a "home court advantage," that the home team wins more than half of all games, in the National Basketball Association. Follow the steps below to conduct a test of significance for whether the sample data support the hypothesis of a home court advantage.

(a) State the null and alternative hypotheses in symbols.
(b) Restate the null and alternative hypotheses in words.

Recall that of the 30 NBA games played from January 3 through January 7, 1992, 18 were won by the home team.

(c) Calculate (by hand) the test statistic.
(d) Calculate the p-value of the test.
(e) Write a one-sentence conclusion, based on the p-value, concerning whether the sample data provide strong evidence of a home-court advantage.
(f) Explain how this test result relates to your analysis in Activity 14-8.

Activity 19-5: Hiring Discrimination

In the case of *Hazelwood School District v. United States* (1977), the U.S. Government sued the City of Hazelwood, a suburb of St. Louis, on the grounds that it discriminated against blacks in its hiring of school teachers. The statistical evidence introduced noted that of the 405 teachers hired in 1972 and 1973 (the years following the passage of the Civil Rights Act), only 15 had been black. The proportion of black teachers living in the county of St. Louis at the time was 15.4% if one includes the city of St. Louis and 5.7% if one does not include the city.

(a) Conduct a significance test to assess whether the proportion of black teachers hired by the school district is statistically significantly less than 15.4% (the percentage of black teachers in the county). Use the .01 significance level. Along with your conclusion, report the null and alternative hypotheses, test statistic, and p-value. (You may use the calculator.)

(b) Conduct a significance test to assess whether the proportion of black teachers hired by the school district is statistically significantly less than 5.7% (the percentage of black teachers in the county if one excludes the city of St. Louis). Again use the .01 significance level and report the null and alternative hypotheses, test statistic, and p-value along with your conclusion.

(c) Write a few sentences comparing and contrasting the conclusions of these tests with regard to the issue of whether the Hazelwood School District was practicing discrimination.

Activity 19-6: Lady Tasting Tea

A famous hypothetical example of statistical inference involves a woman who claims that when presented with a cup of tea and milk, she can distinguish more often than not whether it was the tea or the milk that had been poured first. Suppose that you present her with 100 cups of tea and milk which you have prepared (so that you know for each cup whether it was the tea or the milk that was poured first). For each cup you ask her to identify which was poured first.

(a) Determine how many correct identifications she must make in order for the sample result to be statistically significant at the .10 level. (You may either approach this question analytically or by trial and error using the calculator.) Explain how you arrive at your conclusion.

(b) Repeat (a) for the .05 significance level.

(c) Repeat (a) for the .01 significance level.

Activity 19-7: ESP Testing *(cont.)*

Consider yet again the ESP test that you analyzed in Activities 19-1 and 19-2.

(a) For a test with $n = 100$ cards, determine the sample proportion that a subject would have to identify correctly in order for the sample result to be statistically significant at the .05 level. (You may either approach this question analytically or by trial and error using the calculator.) Explain how you arrive at your conclusion.

(b) Repeat (a) for a test with $n = 400$ cards.

(c) With which sample size ($n = 100$ or $n = 400$ cards) does the subject have to identify a higher percentage correctly in order for the sample result to be statistically significant at the .05 level?

Activity 19-8: Television Magic *(cont.)*

Recall from Activity 17-10 that a sample of *Nick at Nite* television viewers produced 810,000 votes for Samantha and 614,000 votes for Jeannie as having more magic powers. Conduct the appropriate test of significance to assess the extent to which these sample data provide evidence that more than half of the population of all *Nick at Nite* viewers favors Samantha over Jeannie. Report the null and alternative hypotheses, the test statistic, and the p-value of the test. Also indicate whether the sample result is statistically significant at the .001 significance level.

Activity 19-9: Marriage Ages *(cont.)*

Reconsider Activity 17-13, in which you analyzed sample data and found the sample proportion of marriages in which the bride was younger than the groom. Conduct a test of significance to address whether the sample data support the theory that the bride is younger than the groom in more than half of all the marriages in the county. Report the details of the test and write a short paragraph describing and explaining your findings.

Activity 19-10: Veterans' Marital Problems *(cont.)*

Refer to the study mentioned in Activity 18-12. U.S. Census figures indicate that among all American men aged 30–44 in 1985, 27% had been divorced at least once. Conduct a test of significance to assess whether the sample data from the study provide strong evidence that the divorce rate among Vietnam veterans is higher than 27%. Report the null and alternative hypotheses (identifying whatever symbols you introduce) and the test statistic and p-value. Also write a one-sentence conclusion and explain why the conclusion follows from the test results.

Activity 19-11: Jury Representativeness *(cont.)*

Refer to the legal case of *Swain v. Alabama* (1965) mentioned in Activity 18-13. Treat the 1050 grand jurors as a simple random sample from the population of all potential grand jurors, and perform the appropriate test of whether the data provide evidence that the proportion of blacks among *all* potential grand jurors is less than .25. Again be sure to state the null and alternative hypotheses and to calculate the test statistic and p-value. Also write a one-sentence conclusion regarding the question of interest. Finally, explain precisely and concisely what the p-value of the test represents (as if you were explaining it to a judge hearing the case).

Activity 19-12: Dentists' Surveys

Suppose that a survey of dentists finds that 60% (3 out of every 5) of the dentists sampled recommend a certain toothbrush. Even if we assume that this is a random sample and that the dentists' opinions are sincere, does this provide strong evidence that more than half of all dentists prefer the toothbrush? Perform a test of significance to address this question for each of the sample sizes listed below. (You may use the calculator.) In each case report the p-value of the test and indicate whether the sample result is statistically significant at the .10 level, at the .05 level, at the .01 level, and at the .001 level.

(a) $n = 25$
(b) $n = 50$
(c) $n = 100$
(d) $n = 500$

WRAP-UP

This topic has introduced you to the very important technique of *tests of significance* by asking you to understand the reasoning process underlying the procedure. You have also encountered the formal structure of tests of significance and explored further the concept of *statistical significance*.

The next topic will continue your study of significance testing by asking you to examine more properties of significance, the connection between confidence and significance, and some ways in which tests of significance are improperly applied.

TESTS OF SIGNIFICANCE II

OVERVIEW

In the last topic, you discovered the fundamental statistical technique of *tests of significance*, concentrating on the reasoning, structure, and interpretation involved with such tests. You will continue to study the principles of significance testing in this topic, focusing on the application of the procedure to problems involving a population proportion. You will also explore some common ways in which significance testing is misapplied in many situations.

OBJECTIVES

- To continue to develop an understanding of the reasoning, structure, and interpretation of tests of significance.
- To investigate and understand the relationship between *two-sided tests* and one-sided tests.
- To discover the duality between confidence intervals and two-sided significance tests.
- To understand the inappropriateness of treating commonly used significance levels as "sacred."
- To recognize the distinction between *practical significance* and statistical significance.
- To continue to appreciate the assumption of random sampling that underlies statistical inference procedures.

- To apply tests of significance to genuine problems and to interpret the results.

1. Do you suspect that the penny spinning results achieved by this class provides strong evidence that a spun penny is more likely to land on one side than the other?

2. Do you suspect that the sample gathered by the class about students' having their own credit cards provides strong evidence that the proportion of credit card holders among all students at this college differs from one-half?

3. If 24,643 of a random sample of 50,000 college students have their own credit card, would you be fairly convinced that less than half of all college students have their own credit card?

4. If 24,643 of a random sample of 50,000 college students have their own credit card, would you be fairly convinced that *much* less than half of all college students have their own credit card?

IN-CLASS ACTIVITIES

Recall from the previous topic that a test of significance concerning a population proportion has the following form:

$$H_0 : \theta = \theta_0$$
$$H_a : \theta < \theta_0, \quad \text{or} \quad H_a : \theta > \theta_0, \quad \text{or} \quad H_a : \theta \neq \theta_0$$
$$\text{Test statistic: } z = \frac{\hat{p} - \theta_0}{\sqrt{\dfrac{\theta_0(1 - \theta_0)}{n}}}$$
p-value: $\Pr(Z < z)$ or $\Pr(Z > z)$ or $2\Pr(Z > |z|)$

Keep these comments in mind as you apply this procedure:

- θ denotes the true population proportion. We do not know its value in practice, so we refer to it by the symbol θ.
- θ_0 denotes the *hypothesized* value of the population proportion. Its value is determined from the problem at hand.
- The alternative hypothesis H_a can take one of three forms, depending on the question being addressed. The alternative hypothesis should be specified by the experimenter prior to collecting the data.
- The *test statistic z* is calculated simply by normalizing the sample proportion $\hat{p}$.
- The *p-value* is found from the test statistic by using the table of standard normal probabilities. How to find the p-value depends on the form of H_a, the alternative hypothesis. The p-value of the test is the probability, if the null hypothesis were true, of obtaining a sample result as extreme as or more extreme than the one actually obtained. Thus, the *smaller* the p-value, the stronger the evidence *against* the null hypothesis.
- If one uses a prespecified significance level α, the test decision is to reject H_0 if (and only if) the p-value of the test is less than or equal to α. In other words, the sample data are statistically significant at level α if (and only if) p-value $\leq \alpha$.
- The assumptions which underlie the validity of this procedure are that the sample taken is a simple random sample (SRS) from the population of interest; and that the sample size is large ($n \geq 30$ as a rule-of-thumb). One should always ask whether these assumptions are satisfied before believing the results of the procedure.

Activity 20-1: Penny Spinning (*cont.*)

Suppose that you spin a penny as described in Activity 17-1 with the intention of determining whether a spun penny is equally likely to land heads or tails; i.e., whether it would land heads 50% of the time in the long run.

(a) Identify in a symbol and in words the *parameter* of interest in this experiment.

(b) Considering the stated goal of the study, is the alternative hypothesis one-sided or two-sided in this case? Explain.

Researchers use a *two-sided* alternative hypothesis when they suspect that a parameter may fall on either side of the hypothesized value. A one-sided test should be used only when one direction is of interest. For example, with ESP testing, one tests whether the subject does *better* than guessing; it makes little sense to test whether a person does worse than he or she would by just guessing.

(c) Specify the null and alternative hypotheses for this experiment, both in symbols and in words.

(d) Suppose that you spin the penny 150 times, obtaining 65 heads and 85 tails. Use the 1-PropZTest command of your calculator to compute the test statistic and p-value of the test.

(e) Use the calculator to compute how the test statistic and p-value would have been different if the 150 spins had produced 85 heads and 65 tails.

(f) In either of these cases (85 of one outcome and 65 of the other), do the sample data provide strong evidence that a spun penny would *not* land heads 50% of the time in the long run? Are the sample results statistically significant at the .05 significance level?

(g) Now suppose that the goal of the experiment is to investigate whether a spun penny tends to land heads *less than* 50% of the time. Restate the null and alternative hypotheses (in symbols).

(h) Use the calculator to determine the test statistic and p-value of the one-sided test, assuming that 65 of the 150 sample spins landed heads. Report the test statistic and p-value of the test, and comment on how they compare to the ones found with the two-sided test.

(i) Still supposing that the goal of the experiment is to investigate whether a spun penny tends to land heads less than 50% of the time, use the calculator to determine the test statistic and p-value of the one-sided test assuming that 85 of the 150 sample spins landed heads. Again report the test statistic and p-value of the test, and comment on how they compare to the ones found with the two-sided test.

(j) For the sample results of question (i), explain why the formal test of significance is unnecessary.

(k) Summarize your findings of this activity in the following table:

Sample result	Alternative hypothesis	Test statistic	p-value
65 heads, 85 tails	$\theta \neq .5$		
85 heads, 65 tails	$\theta \neq .5$		
65 heads, 85 tails	$\theta < .5$		
85 heads, 65 tails	$\theta < .5$		

Activity 20-2: American Moral Decline *(cont.)*

Reconsider the survey described in Activity 18-1 concerning moral decline in America. Recall that 568 of 748 respondents answered affirmatively to the question, "Do you think the United States is in a moral and spiritual decline?"

(a) Use the calculator to produce a 95% confidence interval for θ, the proportion of all adult Americans who would have answered "yes" to the question. Record the interval below.

(b) Does the value .5 fall within this interval?

(c) Use the calculator to perform a two-sided test of whether the sample provides strong evidence that θ differs from one-half. Report the test statistic and p-value; also indicate whether the sample proportion differs significantly from one-half at the $\alpha = .05$ significance level. Record the results in the table below.

(d) Use the calculator to perform a two-sided test of whether the sample provides strong evidence that θ differs from 70%. Again record the results in the table below, and also indicate whether the 95% confidence interval for θ includes the value .7.

(e) Repeat (d) where the value of interest is 75% rather than 70%. Again record the results in the table below, and also indicate whether the 95% confidence interval for θ includes the value .75.

(f) Repeat (d) for the value 78%.

(g) Repeat (d) for the value 80%.

Hypothesized value of θ	Contained in 95% c.i.?	Test statistic	(Two-sided) p-value	Significant at .05 level?
.50				
.70				
.75				
.78				
.80				

(h) Do you notice any connection between whether a 95% confidence interval for θ includes a particular value and whether the sample proportion differs significantly from that particular value at the $\alpha = .05$ level? Explain.

This activity reveals a *duality* between confidence intervals for estimating a parameter and a two-sided test of significance about the value of the parameter. Roughly speaking, if a 95% confidence interval for a parameter does not include a particular value, then a two-sided test of whether the parameter equals that particular value will be statistically significant at the $\alpha = .05$ level.

Confidence intervals and tests of significance are complementary procedures. While tests of significance can establish strong evidence that an effect exists, confidence intervals serve to estimate its magnitude.

Activity 20-3: Advertising Strategies

Suppose that the managers of a company want to decide whether more than half of the people who use the company's product are women, because they are planning to launch a new advertising campaign if they discover that most of their customers are women. They ask you to take a simple random sample of 200 of their customers and to report *only* whether the sample proportion of women is significantly greater (at the $\alpha = .05$ significance level) than one-half.

(a) Suppose that 111 of the customers sampled are women. Determine (using the calculator) the p-value of the appropriate significance test. Report the test statistic and p-value; also indicate whether or not the sample proportion is statistically significantly (at the .05 level) greater than one-half.

(b) Repeat (a), but supposing that 112 of the customers sampled are women.

(c) Repeat (a), but supposing that 124 of the customers sampled are women.

(d) Are the sample results more similar in (a) and (b) or in (b) and (c)?

(e) Would your report to the company be the same in (a) and (b) or in (b) and (c)?

The moral here is that it is unwise to treat standard significance levels as sacred. It is much more informative to consider the p-value of the test and to base one's decision on it. There is no sharp border between "significant" and "insignificant," only increasingly strong evidence as the p-value decreases. If you were responsible for analyzing the data and reporting to the company's managers, you would provide them with the best information by giving them the sample proportion and the p-value of the test (along with an explanation of what the p-value means). On the other hand, if you were one of the managers, you would serve your company best by asking for that information instead of just a statement of significance.

Activity 20-4: Tax Return Errors (*cont.*)

Suppose that the IRS wants to decide if the proportion of U.S. taxpayers who make an arithmetic error on their personal income tax return differs significantly from 30%. Suppose that they take a simple random sample of 50,000 returns and find that 15,313 of these contain arithmetic errors.

(a) What is the sample proportion of returns which contain errors?

(b) Does this sample proportion seem to differ significantly from 30% in a practical sense?

(c) Conduct (using the calculator) the appropriate significance test to determine if the sample proportion *differs* statistically significantly (at the .01 level, say) from 30%. Record the null and alternative hypotheses, test statistic, and p-value.

(d) Find (using the calculator) a 99% confidence interval for the actual proportion of *all* tax returns which contain arithmetic errors.

(e) Does this interval contain the value .3?

(f) Do the values in the interval differ substantially (in a practical sense) from .3?

This example illustrates that *statistical* significance is not the same thing as *practical* significance. A statistically significant result is simply one which is unlikely to have occurred by chance; that does not necessarily mean that the result is substantial or important in a practical sense. While the IRS would have strong reason to believe that the true proportion of returns containing errors *does* indeed *differ from* 30%, that true proportion is actually *very close to* 30%. Especially when one works with very large samples, an unimportant result can be statistically significant nonetheless. Confidence intervals are useful for estimating the size of the effect involved and should be used in conjunction with significance tests.

Activity 20-5: Elvis and Alf (*cont.*)

Recall from Activity 12-1 that in 1936 the *Literary Digest* received 2.4 million responses to their survey, with 57% of the sample indicating support for Alf Landon over Franklin Roosevelt in the upcoming election.

(a) Use the calculator to conduct the appropriate test of significance to determine if the sample proportion expressing support for Landon is statistically significantly greater than one-half. Record the test statistic and p-value of the test; also indicate whether the sample result is significant at the .001 level.

(b) Use the calculator to find a 99.9% confidence interval for θ, the proportion of all registered voters at the time who intended to vote for Landon.

(c) Explain why the test of significance led to such an erroneous conclusion in this case.

This activity should remind you that all of the inference techniques we study depend upon the data having been collected by random sampling. When the method of data collection is biased, the results of inference procedures are not only invalid but potentially misleading as well.

HOMEWORK ACTIVITIES

Activity 20-6: College Students' Credit (*cont.*)

Reconsider Activity 18-4 in which you analyzed sample data on whether or not college students have their own credit card. Let θ represent the proportion of all students at this college who have their own credit card.

Suppose that you want to test whether the sample data provide strong evidence that θ differs from one-half.

(a) State the null and alternative hypotheses in both symbols and words.
(b) Calculate (by hand) the test statistic.
(c) Calculate the p-value of the test.
(d) Write a sentence or two interpreting this test result and explaining your conclusion about the question of whether the proportion of students at this college who have a credit card differs from one-half.
(e) Does the sample proportion differ significantly from one-half at the $\alpha =$ .10 significance level?
(f) Find (by hand) a 90% confidence interval for θ.
(g) Does the 90% confidence interval for θ contain the value .5? Explain how this question relates to question (e).
(h) Comment on whether the assumptions of the inference procedures seem to be satisfied in this study.

Activity 20-7: Penny Spinning *(cont.)*

Reconsider the penny spinning experiment conducted in Activity 17-1. Use the sample results collected in class to test whether the data provide evidence that θ, the long-term proportion of heads that would result from *all* hypothetical penny spins, *differs from* one-half. You may conduct the test by hand or by using the calculator, but be sure to report the null and alternative hypotheses, the test statistic, and the p-value. Also write a brief conclusion.

Activity 20-8: Cat Households *(cont.)*

Recall from Activity 17-11 that a survey of 40,000 American households in 1987 found that 30.5% of the households in the sample owned a pet cat.

(a) Is this sample proportion statistically significantly *less than* one-third at the .001 significance level? Support your answer with the details of the significance test.
(b) Form a 99% confidence interval to estimate the proportion of all American households that owned a pet cat in 1987.
(c) Comment on whether the sample data suggest that the proportion of cat owners among all American households is significantly less than one-third in a practical sense.

Activity 20-9: Voter Dissatisfaction with Congress *(cont.)*

Recall from Activity 18-5 that a nationwide telephone poll of 1007 adults in October of 1994 found 43% agreeing that "most U.S. representatives deserve to be re-elected."

(a) Conduct (by hand) the appropriate test of significance to assess whether this sample information provides strong evidence that the proportion of all adult Americans who agree with this proposition differs from one-half. Report the null and alternative hypotheses, and calculate (by hand) the test statistic and p-value of the test. Also state your conclusion and indicate whether the sample proportion differs significantly from one-half at the .01 significance level.

(b) Based solely on your finding in (a), would you expect a 99% confidence interval for the population proportion to contain the value .5? Explain.

Activity 20-10: Telephone Books *(cont.)*

Recall from Activity 18-7 the sample data that you collected from telephone books. Using your sample data, perform a test of significance to address whether the sample data support the theory that less than half of all of the telephone book's individual listings carry female names. Report the details of the test and write a short paragraph describing your findings and explaining how your conclusions follow from the test results.

Activity 20-11: Female Senators *(cont.)*

Recall from Activity 18-3 that the 1994 U.S. Senate consists of 7 women and 93 men.

(a) Treat these figures as sample data and calculate the test statistic for the significance test of whether the population proportion of women is less than 50%.

(b) Use the test statistic and the table of standard normal probabilities to calculate the p-value of the test.

(c) If the goal here is to decide whether women comprise less than half of the entire U.S. Senate of 1994, does this test of significance have any meaning? Explain.

Activity 20-12: Taste Testing Experiments

Suppose that you conduct a taste-testing experiment to assess whether people tend to prefer one brand of cola more often than another when the two are compared head-to-head. You are not particularly interested in which brand is preferred; you want to study whether either is preferred over the other. You plan to present 250 people with a cup of cola A and a cup of cola B.

(a) Identify in a symbol and in words the *parameter* of interest in this experiment.

(b) Considering the stated goal of the study, is the alternative hypothesis one-sided or two-sided in this case? Explain.

(c) Specify the null and alternative hypotheses for this experiment, both in symbols and in words.

(d) Suppose that 135 of the 250 subjects in the sample prefer cola A to cola B. Calculate (by hand) the test statistic and p-value of the test.

(e) How would the test statistic and p-value have been different if 115 of the 250 subjects preferred cola A to cola B? Perform the calculations to support your answer.

(f) In either of the cases in (d) and (e), do the sample data provide strong evidence that people tend to prefer one brand of cola more often than the other? Are the sample results statistically significant at the .05 significance level?

(g) Now suppose that the goal of the experiment is to establish that people tend to prefer brand A more often than brand B. Restate the null and alternative hypotheses (in symbols).

(h) Use the calculator to determine the test statistic and p-value of the one-sided test assuming that 135 of the 250 sample subjects preferred brand A over brand B. Report the test statistic and p-value of the test, and comment on how they compare to the ones found with the two-sided test.

Activity 20-13: Home Court Advantage *(cont.)*

Consider again the sample information of Activity 19-4 that the home team won 18 of a sample of 30 games played in the National Basketball Association.

(a) Report the p-value of the significance test of whether the sample data support the hypothesis of a home court advantage.

(b) Find (by hand) a 90% confidence interval for the proportion of all NBA games won by the home team.

(c) Does this sample information convince you that there is a home court advantage in the NBA? Does it convince you that there is *not* a home-court advantage in the NBA? Explain.

Activity 20-14: Age and Political Ideology *(cont.)*

Refer to Activity 17-14, in which you constructed a confidence interval for the proportion of all American adults who identify themselves as political moderates.

(a) Conduct (either by hand or with the calculator) a test of significance of whether the sample data provide evidence that the proportion of all American adults who identify themselves as political moderates differs from one-half. Report the null and alternative hypotheses, the test statistic, and p-value. Also write a one-sentence conclusion.
(b) Comment on how the conclusion of this test relates to the confidence interval that you found in Activity 17-14.

Activity 20-15: College Students' Credit (*cont.*)

Suppose that you want to study the question of college students' having their own credit cards on the national level, so you take a random sample of 50,000 college students. Suppose you find that 24,643 of these students have their own credit card.

(a) Does this sample information provide strong evidence that less than half of all American college students have their own credit card? Support your answer with appropriate calculations and explanations.
(b) Does this sample information provide evidence that the proportion of all American college students who have their own credit card is very much less than one-half? Again support your answer with appropriate calculations and explanations.

Activity 20-16: Phone-In Polling

Suppose that a local 12:00 noon news show asks viewers to call in to say whether they support a bill that would extend benefits for unemployed Americans. If 700 people call in and 500 say that they support the bill, would you conclude that more than half of *all* local residents support the bill? Write a brief paragraph to support your answer.

WRAP-UP

In addition to providing you with more experience conducting and interpreting *tests of significance*, this topic has led you to consider some important ideas related to significance testing. You have explored the properties of *two-sided* as opposed to one-sided tests, investigated the unwise use of fixed significance levels, and discovered the distinction between statistical significance and *practical significance*.

In the next unit you will apply the inference techniques of confidence intervals and significance tests to experiments which involve *comparing* two population proportions. As you proceed to study confidence intervals and tests of significance in a variety of contexts, you will find that the structure, reasoning, and interpretation of these procedures do not change. Only the details of the calculations vary from situation to situation.

Before you study these inference procedures, you will first discover the need for *controlled experiments* in situations where one wants to detect a cause-and-effect relationship between variables.

Inference from Data: Comparisons

DESIGNING EXPERIMENTS

You have studied the idea of random sampling as a fundamental principle by which to gather information about a population. Since then you have discovered formal techniques of statistical inference which enable you to draw conclusions about a population based on analysis of a sample. With this topic you will begun to study another technique of data collection for use when the goal is not to describe a population but to investigate the effects that a variable has on other variables. You will investigate the need for *controlled experiments* and discover some fundamental principles which govern the design and implementation of such experiments.

- To recognize the need for carefully designed *experiments* in order to detect a cause-and-effect relationship between variables.
- To learn to distinguish between *explanatory variables* and *response variables*.
- To appreciate the principle of *control* as the fundamental idea of experimental design.
- To discover some principles commonly used to achieve control in designed experiments.
- To develop the ability to detect violations of these principles in studies.
- To understand the distinction between *controlled experiments* and *observational studies* and the different conclusions that one can draw from each.

- To learn to describe in considerable detail how one might design and conduct experiments on practical issues.

PRELIMINARIES

1. If 95% of the participants in a large SAT coaching program improve their SAT scores after attending the program, would you conclude that the coaching was responsible for the improvement?

2. If recovering heart attack patients who own a pet tend to survive longer than recovering heart attack patients who do not own a pet, would you conclude that owning a pet has therapeutic benefits for heart attack patients?

3. Do you think that eating SmartFood™ popcorn really makes a person smarter?

4. Take a guess as to the mean IQ of a sample of people who claim to have had intense experiences with unidentified flying objects (UFO's).

5. Do you suspect that people who claim to have had intense experiences with UFO's tend to have higher or lower IQ's than other people, or do you think that there is no difference in their IQ's on the average?

6. If high school students who study a foreign language tend to perform better on the verbal portion of the SAT than students who do not study a foreign language, does that establish that studying a foreign language improves one's verbal skills?

7. If states which have capital punishment tend to have lower homicide rates than states without capital punishment, would you attribute that difference to the deterrence effect of the death penalty?

8. If states which have capital punishment have similar homicide rates to states without capital punishment, would you conclude that the death penalty has no deterrence effect?

In-Class Activities

Activity 21-1: SAT Coaching

Suppose that you want to study whether an SAT coaching program actually helps students to score higher on the SAT's, so you gather data on a random sample of students who have attended the program. Suppose you find that 95% of the sample scored higher on the SAT's after attending the program than before attending the program. Moreover, suppose you calculate that the sample mean of the improvements in SAT scores was a substantial 120 points.

(a) Explain why you cannot legitimately conclude that the SAT coaching program *caused* these students to improve on the test. Suggest some other explanations for their improvement.

The SAT coaching study illustrates the need for a *controlled experiment* to allow one to draw meaningful conclusions about one variable *causing* another to respond in a certain way. The fundamental principle of experimental design is *control*. An experimenter tries to *control* for possible effects of extraneous variables so that any differences observed in one variable of interest can be attributed directly to the other variable of interest.

The variable whose effect one wants to study is called the *explanatory variable* (or *independent* variable), while the variable which one suspects to be affected is known as the *response variable* (or *dependent* variable).

The counterpart to a controlled experiment is an *observational study* in which one passively records information without actively intervening in the process. As you found when you discovered the distinction between correlation and causation in Topic 7 (think of televisions and life expectancies), one cannot infer causal relationships from an observational study since the possible effects of *confounding variables* are not controlled.

(b) Identify the explanatory and response variables in the SAT coaching study.

(c) Is the SAT coaching study as described above a controlled experiment or an observational study? Explain.

Experimenters use many techniques in their efforts to establish control. One fundamental principle of control is *comparison*. One important flaw in that SAT coaching study is that it lacks a *control group* with which to compare the results of the group that attends the coaching program.

Activity 21-2: Pet Therapy

Suppose that you want to study whether pets provide a therapeutic benefit for their owners. Specifically, you decide to investigate whether heart attack patients who own a pet tend to recover more often than those who do not. You randomly select a sample of heart attack patients from a large hospital and follow them for one year. You then compare the sample proportions who have survived and find that 92% of those with pets are still alive while only 64% of those without pets have survived.

(a) Identify the explanatory and response variables in this study.

(b) Is this study a controlled experiment or an observational study? Explain.

This pet therapy study points out that analyzing a comparison group does not guarantee that one will be able to draw cause-and-effect conclusions. A critical flaw in the design of the pet therapy study is that subjects naturally decide for themselves whether or not to own a pet. Thus, those who opt to own a pet may do so because they are in better health and therefore more likely to survive even if the pet is of no benefit at all.

Experimenters try to assign subjects to groups in such a way that confounding variables tend to balance out between the two groups. A second principle of control which provides a simple but effective way to achieve this is *randomization*. By randomly assigning subjects to different treatment groups, experimenters ensure that hidden confounding variables will balance out between/among the groups in the long run.

Activity 21-3: Vitamin C and Cold Resistance

Suppose that you want to design a study to examine whether taking regular doses of vitamin C increases one's resistance to catching the common cold. You form two groups—subjects in one group receive regular doses of vitamin C while those in the other (control) group are not given vitamin C. You assign subjects at random to one of these two groups. Suppose that you then monitor the health of these subjects over the course of a winter and find that 46% of the vitamin C group resisted a cold while only 14% of the control group resisted a cold.

(a) Identify the explanatory and response variables in this study.

(b) Is this vitamin C study a controlled experiment or an observational study? Explain.

(c) Identify some of the confounding variables which affect an individual's resistance to a cold that the random assignment of subjects should balance out.

The design of the vitamin C study could be improved in one respect. There is one subtle confounding variable that the randomization cannot balance out because this variable is a direct result of the group to which a subject is assigned. The very fact that subjects in the vitamin C group realize that they are being given something that researchers expect to improve their health may cause them to remain healthier than subjects who are not given any treatment. This phenomenon has been detected in many circumstances and is known as the *placebo effect*.

Experimenters control for this confounding variable by administering a placebo ("sugar pill") to those subjects in the control group. This third principle of control is called *blindness* since the subjects are naturally not told whether they receive vitamin C or the placebo. When possible, experiments should be *double-blind* in that the person responsible for evaluating the subjects should also be unaware of which subjects receive which treatment. In this way the evaluator's judgment is not influenced (consciously or subconsciously) by any hidden biases.

Activity 21-4: Pregnancy, AZT, and HIV (*cont.*)

Recall from Activity 11-3 that a study was performed to investigate whether the drug AZT reduces the risk of an HIV-positive pregnant woman giving birth to an HIV-positive baby.

(a) Identify the explanatory and response variables in this study. Are these measurement or categorical variables? If they are categorical, are they also binary?

(b) Explain how the study could be designed to make use of the principle of *comparison*.

(c) Explain how the study could be designed to incorporate the principle of *randomization*.

(d) Explain how the study could be designed to take into account the principle of *blindness*.

In the context of medical studies, controlled experiments such as the AZT study which randomly assign subjects to a treatment are called *clinical trials*. Many medical issues do not lend themselves to such studies, however, so researchers must resort to observational studies. Three types of observational studies are often used in medical contexts:

- *Case-control studies*, in which one starts with samples of subjects who do and who do not have the disease and then looks back into their histories to see which have used and which have not used a certain treatment or did or did not have some condition;
- *Cohort studies*, in which one starts with samples of subjects who do and who do not use or have the treatment or condition, and then follows them into the future to see which do and which do not develop the disease;
- *Cross-sectional studies*, in which one simply takes a sample of subjects and classifies them according to both variables (have disease or not, use treatment or not).

While controlled experiments are the only way to establish a causal relationship between variables, observational studies can also provide researchers with important information.

Activity 21-5: Smoking and Lung Cancer

(a) Explain how one could, in principle, design a controlled experiment to assess whether smoking causes lung cancer. Also indicate why conducting such an experiment would be morally unethical.

(b) What kind of observational study are researchers conducting if they take a random sample of 500 people who suffer from lung cancer and a random sample of 500 people who do not suffer from lung cancer and then compare the proportions of smokers in each group?

(c) What kind of observational study are researchers conducting if they take a random sample of 500 smokers and a random sample of 500 non-smokers, follow them for 20 years, and compare the proportions who develop lung cancer in each group?

Activity 21-6: SmartFood Popcorn

Describe (in as much detail as possible) how you might design an experiment to assess whether eating SmartFood™ brand popcorn really makes people smarter. You will have to describe how you would measure "smartness" as part of the experimental design. Try to adhere to the three principles of control that we listed above.

HOMEWORK ACTIVITIES

Activity 21-7: UFO Sighters' Personalities

In a 1993 study, researchers took a sample of people who claimed to have had an intense experience with an unidentified flying object (UFO) and a sample of people who did not claim to have had such an experience. They then compared the two groups on a wide variety of variables, including IQ. The sample mean IQ of the UFO group was 101.6 and that of the control group was 100.6. Is this study a controlled experiment or an observational study? If it is an observational study, what kind is it? Explain your answers.

Activity 21-8: Mozart Music

Researchers in a 1993 study investigated the effect of listening to music by Mozart before taking an IQ test. Subjects were randomly assigned to one of three groups and would either listen to Mozart music, be told to relax, or be given no instructions. The sample mean IQ in the Mozart group was 119, in the relax group was 111, and in the silent group was 110.

(a) Is this study a controlled experiment or an observational study? If it is an observational study, what kind is it? Explain your answers.
(b) Identify the explanatory variable and the response variable.
(c) Classify each variable as a categorical or measurement variable.

Activity 21-9: Language Skills

Students who study a foreign language in high school tend to perform better on the verbal portion of the SAT than students who do not study a foreign language. Can you conclude from these studies that the study of a foreign language causes students to improve their verbal skills? Explain.

Activity 21-10: Capital Punishment

Suppose that you want to study whether the death penalty acts as a deterrent against homicide, so you compare the homicide rates between states that have the death penalty and states that do not.

(a) Is this a controlled experiment? Explain.
(b) If you find a large difference in the homicide rates between these two types of states, can you attribute that difference to the deterrence effect of the death penalty? Explain.
(c) If you find no difference in the homicide rates between the two types of states, can you conclude that the death penalty has no deterrence effect? Explain.

Activity 21-11: Literature for Parolees

In a recent study, 32 convicts were given a course in great works of literature. To be accepted for the program the convicts had to be literate and to convince a judge of their intention to reform. After 30 months of parole only 6 of these 32 had committed another crime. This group's performance was compared against a similar group of 40 parolees who were not given the literature course; 18 of these 40 had committed a new crime after 30 months.

(a) What proportion of the literature group committed a crime within 30 months of release? What proportion of this group did not commit a crime?
(b) What proportion of the control group committed a crime within 30 months of release? What proportion of this group did not commit a crime?
(c) Construct a segmented bar graph to compare these conditional distributions of crime commission between the literature and control groups.
(d) Which fundamental principles of control does this experiment lack? Comment on how this lack hinders the conclusion of a cause-and-effect relationship in this case.

Activity 21-12: Gun Control Legislation

(a) Suppose that a nation passes a strict gun control measure and finds five years later that the national homicide rate has increased. Can you conclude that the passage of the gun control measure *caused* the homicide rate to increase? Explain.
(b) Would your answer to (a) differ if the homicide rate had actually *decreased* and you were asked to conclude that the passage of the gun control measure *caused* the homicide rate to decrease? Explain.

Activity 21-13: Baldness and Heart Disease (*cont.*)

Reconsider the study mentioned in Activity 11-14, where researchers took a sample of male heart attack patients and a sample of men who have not suffered a heart attack and compared baldness ratings between the two groups. Their goal was to determine if baldness has an effect on one's likelihood of experiencing a heart attack.

(a) Identify the explanatory and response variables.
(b) Is this study a controlled experiment or an observational study? If it is an observational study, what kind is it? Explain your answers.

Activity 21-14: Pet Therapy (*cont.*)

Reconsider the study described in Activity 21-2. Explain how you could (in principle) design a controlled experiment to investigate the proposition that owning a pet has therapeutic benefits for heart attack patients.

Activity 21-15: Assessing Calculus Reform

Many colleges and universities in the 1990's have developed "calculus reform" courses which substantially alter the way that calculus is taught. The goal is that the reform courses help students to understand fundamental calculus concepts better than traditionally taught courses do.

(a) If you simply compare scores on a standardized calculus test between students in colleges that teach a reform course and students in colleges that teach a traditional course, would you be able to conclude that any differences you might find are attributable to the teaching style (reform or traditional)?
(b) Describe how you might design an experiment to assess whether the goal is being met. Be sure that your experimental design incorporates aspects of comparison, randomization, blindness, and double-blindness. Also explain the need for each of these aspects in your design.

Activity 21-16: Subliminal Messages

Explain in as much detail as possible how you might design and conduct an experiment to assess whether listening to audiotapes with recorded subliminal messages actually helps people to lose weight.

Activity 21-17: Experiments of Personal Interest

Think of a situation in which you would be interested in determining whether a cause-and-effect relationship exists between two variables. Describe in as much detail as possible how you might design and conduct a controlled experiment to investigate the situation.

WRAP-UP

This topic has introduced you to the principles of designing *controlled experiments*. You have explored the limitations of *observational studies* with regard to establishing causal relationships between variables. You have also learned that *control* is the guiding principle of *experimental design* and discovered the principles of *comparison*, *randomization*, and *blindness* as specific techniques for achieving control.

In the next two topics you will investigate how inference procedures such as tests of significance and confidence intervals enable one to draw conclusions about the results of experiments. As you have done in the context of sampling from a population, you will first study inferential methods for comparing two *proportions*.

COMPARING TWO PROPORTIONS I

In the previous topic you discovered the necessity of controlled experiments in scientific investigation and found that randomization plays an important role in the design of experiments. Just as it did with regard to sampling from a population, this random component enables us to draw inferences from the results of experiments. In this topic you will investigate the role of tests of significance in the context of *comparing* proportions between two experimental groups. You will see that the basic reasoning, structure, and interpretation of these tests are the same as in the case of inference about a single population proportion.

- To understand the reasoning behind tests of significance for comparing sample proportions between two experimental groups.
- To explore properties of tests of significance for comparing two proportions.
- To gain experience with applying these significance tests to actual data and in interpreting the results.

PRELIMINARIES

1. Suppose that 10 patients are randomly assigned to receive a new medical treatment while another 10 are randomly assigned to receive the old standard treatment. If 7 patients in the "new" group and 5 patients in the "old" group recover, would you be fairly convinced that the new treatment is superior to the old?

2. Suppose that 10 patients are randomly assigned to receive a new medical treatment while another 10 are randomly assigned to receive the old standard treatment. If 9 patients in the "new" group and 3 patients in the "old" group recover, would you be fairly convinced that the new treatment is superior to the old?

3. Take a guess concerning the proportion of Americans who would respond in the affirmative to the question, "Do you think the United States should forbid public speeches in favor of communism?"

4. Suppose that one group of subjects is asked the question in #3 above, while another group is asked, "Do you think the United States should allow public speeches in favor of communism?". Would you expect these groups to differ in terms of their opposition to communist speeches? If so, which group do you think would more likely oppose communist speeches?

5. Take a guess as to the proportion of students at your college who would favor a policy to eliminate smoking from all buildings on campus.

6. In Activity 22-1 you will be comparing a new treatment with an old standard treatment by exploring hypothetical medical recovery rates. In or-

der to obtain a better understanding of this situation, you need to run a simulation of 1,000 random assignments of patients to treatments. The program that runs this simulation on your calculator is named MEDICAL.83p. Download this program into your calculator. Enter 1,000 for the number of assignments at the prompt and press ENTER. This simulation will take approximately 25 minutes. The data are stored in a list named MED. The information obtained from this simulation will be used in Activity 22-1 parts (h) − (j).

IN-CLASS ACTIVITIES

Activity 22-1: Hypothetical Medical Recovery Rates

Suppose that medical researchers want to compare a new treatment (N) to the old standard treatment (O). They design an experiment in which they take a sample of patients suffering from the disease and randomly assign half to receive treatment N and half to receive treatment O. They then see how many in each group experience a satisfactory recovery. They hope, of course, that the data will provide strong evidence that the proportion of *all* patients who would recover from the new treatment exceeds that from the old treatment. Suppose that the experiment involves 20 patients and that 10 are assigned at random to each treatment.

(a) Suppose that 7 patients receiving treatment N and 5 receiving treatment O recover. What are the sample proportions of recovery in each group? Which treatment produced the higher proportion of recovery? In light of these data, how convinced (intuitively) would you be that this difference in recovery rates is not just due to chance; i.e., that treatment N is really superior to treatment O? Explain.

In keeping with the reasoning of significance tests, we will ask how likely the sample results would have been if in fact the two treatments were equally effective. One way to analyze this question is to assume that 12 of the patients in the study would have recovered regardless of which treatment they had been given. We can then *simulate* the process of assigning them at random to the two treatment groups, observing how often we obtain sample results as extreme as the actual.

(b) Take 20 cards to represent the patients in the study. Mark 12 of them as recoveries and 8 as deaths. (If you use playing cards, you might use red cards as recoveries and black as deaths.) Shuffle them and deal out 10 at random to represent those assigned to treatment N. How many "recoveries" and "deaths" did your random assignment produce for the treatment N group?

(c) Is your simulated result as extreme as the actual result? In other words, are there seven or more recoveries in your simulated treatment N group?

(d) Repeat this process a total of five times. For each repetition record the number of recoveries and deaths assigned by chance to the treatment N group in the table:

Repetition	1	2	3	4	5
Treatment N recoveries					
Treatment N deaths					

(e) Now combine your simulated results with those of the rest of the class. Create a dotplot below of the number of simulated experiments which produced the indicated number of *recoveries* in treatment N. Notice that the dotplot marks with an arrow the actual sample result of seven recoveries with treatment N.

```
--------+---------+---------+------+--+---------+-------- N recoveries
       3.0       4.5       6.0     ↑ 7.5       9.0       10.5
```

(f) Use the dotplot to create a tally below of the number of simulated experiments which produced the indicated number of recoveries and deaths for treatment N. Notice that outcomes as extreme as the actual sample result are boldfaced.

Treatment N recoveries	2	3	4	5	6	**7**	**8**	**9**	**10**
Treatment N deaths	8	7	6	5	4	**3**	**2**	**1**	**0**
Number of simulated experiments									

(g) In how many and what proportion of these simulated experiments does the number of recoveries assigned by chance to treatment N equal seven or more? Is it very unusual for treatment N to have seven or more recoveries even if it is equally effective with treatment O?

(h) Now use the results of your calculator simulation of 1,000 random assignments of patients to treatments. The MED list is set up so that the first entry is the number of assignments with zero treatment N deaths, the second entry is the number of assignments with one treatment N death, and so forth. Enter the results of these 1,000 simulated assignments in the following table; notice again that outcomes as extreme as the actual sample result are boldfaced.

Treatment N recoveries	2	3	4	5	6	**7**	**8**	**9**	**10**
Treatment N deaths	8	7	6	5	4	**3**	**2**	**1**	**0**
Number of simulated experiments									

(i) In how many and what proportion of these 1,000 simulated experiments does the number of recoveries assigned by chance to treatment N equal seven or more? Based on this more extensive simulation, would you say that it is very unusual for treatment N to have seven or more recoveries even if it is equally effective with treatment O? Explain.

(j) Suppose now that the actual experiment had resulted in 9 recoveries and 1 death with treatment N, and 3 recoveries and 7 deaths with treatment O. Would this sample result constitute strong evidence that treatment N is superior to treatment O? Explain your answer based on the simulations performed above.

Rather than rely on simulations, we can use a formal test of significance to assess how likely a sample result would have occurred by chance alone. The reasoning and interpretation of these tests are the same as when you applied them to a single population proportion. Now we are interested in comparing population proportions between two groups; denote these population proportions by θ_1 and θ_2.

The null hypothesis asserts that θ_1 and θ_2 are equal. The alternative hypothesis can again take one of three forms depending on the researchers' conjecture about θ_1 and θ_2 prior to conducting the study. The test statistic is calculated by comparing the two sample proportions and standardizing appropriately. The p-value is again found from the standard normal table. One interprets this p-value as before: it represents the probability of having gotten a sample result as extreme as the actual result if the null hypothesis were true. Thus, the smaller the p-value, the stronger the evidence against the null hypothesis.

The details of this test procedure are summarized below:

Significance test of equality of θ_1 and θ_2:

$$H_0 : \theta_1 = \theta_2$$
$$H_a : \theta_1 < \theta_2, \quad \text{or} \quad H_a : \theta_1 > \theta_2, \quad \text{or} \quad H_a : \theta_1 \neq \theta_2$$

Test statistic: $z = \dfrac{\hat{p}_1 - \hat{p}_2}{\sqrt{\hat{p}_c(1 - \hat{p}_c)\left(\dfrac{1}{n_1} + \dfrac{1}{n_2}\right)}}$

p-value $= \Pr(Z < z)$ or $\Pr(Z > z)$ or $2\Pr(Z > |z|)$

---

where n_1 and n_2 are the respective sample sizes from the two groups, $\hat{p}_1$ and $\hat{p}_2$ are the respective *sample* proportions, $\hat{p}_c$ denotes the *combined* sample proportion (i.e., the proportion of "successes" if the two samples were pooled together as one big sample), and Z represents the familiar standard normal distribution.

Assumptions:

1. The two samples are independently selected simple random samples from the populations of interest.
2. Both sample sizes are large ($n_1 \geq 30$ and $n_2 \geq 30$ as a rule-of-thumb).

Activity 22-2: Pregnancy, AZT, and HIV (*cont.*)

Recall from Activity 11-3 that medical experimenters randomly assigned 164 pregnant, HIV-positive women to receive the drug AZT during pregnancy, while another 160 such women were randomly assigned to a control group which received a placebo. Of those in the AZT group, 13 had babies who tested HIV-positive, compared to 40 HIV-positive babies in the placebo group.

(a) Let θ_{AZT} denote the proportion of *all* potential AZT-takers who would have HIV-positive babies and θ_{plac} denote the proportion of *all* potential placebo-takers who would have HIV-positive babies. Write (using these symbols) the appropriate null and alternative hypotheses for testing the researchers' conjecture that AZT would prove beneficial for these patients.

(b) Calculate (by hand) the sample proportions of HIV-positive babies in the two experimental groups. Record these along with the symbols used to represent them.

(c) A total of how many women were subjects in this study? How many had HIV-positive babies? Use this information to determine $\hat{p}_c$, the combined sample proportion who had HIV-positive babies.

(d) Use your answers to (b) and (c) to calculate the test statistic for testing the hypotheses of (a).

(e) Based on this test statistic, use the table of standard normal probabilities to calculate the p-value of the test.

(f) If AZT and the placebo were equally effective in preventing babies from having HIV, what would be the probability of chance alone assigning 40 HIV-positive babies to the placebo group and only 13 to the AZT group? Is this outcome very unlikely to occur by chance alone?

(g) Based on the p-value, write a one-sentence conclusion about the extent to which the sample data support the researchers' conjecture.

Activity 22-3: Hypothetical Medical Recovery Rates (*cont.*)

Suppose that a medical experiment randomly assigns patients to receive either an "old" or a "new" treatment for a certain condition. Suppose that 60% of those patients receiving the old treatment recover and that 70% of those in the new treatment group recover.

(a) Before evaluating whether this difference in sample proportions of recovery is statistically significant, what further information about the study would you ask for? (Ignore design issues for now; suppose that it is a well-designed controlled experiment.)

(b) Describe a circumstance in which these sample proportions of recovery would not convince you at all that the new treatment is superior to the old.

(c) Describe a circumstance in which these sample proportions of recovery would strongly convince you that the new treatment is superior to the old.

(d) Use the 2-PropZTest feature located in the STAT TESTS menu to determine the p-value of the appropriate significance test for the different sample sizes listed in the table below. In each case assume that 60% receiving the old treatment recovers and 70% receiving the new treatment recovers. Also record whether or not the difference in sample proportions is statistically significant at the .10, .05, and .01 levels.

"Old" treatment sample size	"New" treatment sample size	(One-sided) p-value	Significant at .10?	Significant at .05?	Significant at .01?
50	50				
100	100				
200	200				
500	500				

(e) Write a sentence or two in which you answer (b) and (c) again in light of this investigation.

HOMEWORK ACTIVITIES

Activity 22-4: Literature for Parolees (*cont.*)

Consider the study described in Activity 21-11 concerning the program to enroll convicts in a literature course in the hope that they would be less likely to commit a crime after their release. Of the 32 convicts given the course, only 6 committed a crime within 30 months of release. Of 40 similar parolees not given the literature course, 18 committed a crime in that period.

To assess how likely it is that these sample results would have occurred by chance if there were no difference in the two groups' effectiveness, consider the histogram below. These data are the result of 1,000 repetitions of the simulated random assignment of convicts to groups, assuming that the 24 (i.e., 6+18) who committed a crime would have done so regardless of the group to which they were assigned. The histogram displays the number of crime-committers assigned by chance to the group which took the literature course.

litcrime   *N* = 1000

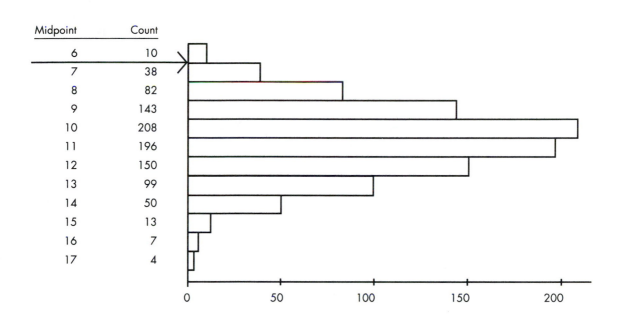

Midpoint	Count
6	10
7	38
8	82
9	143
10	208
11	196
12	150
13	99
14	50
15	13
16	7
17	4

(a) In how many of these 1,000 simulated random assignments were the re-sults as favorable to the literature program as the actual results were? Explain what "as favorable" means in this case. What proportion of the 1,000 repetitions were as favorable to the program as the actual results?

(b) Conduct (by hand) the appropriate test of significance to assess whether the proportion of crime-committers is statistically significantly smaller among those who take the course than among those who do not. Report the null and alternative hypotheses, and show the details of your cal-culations of the test statistic and p-value.

(c) Is the p-value close to the proportion that you found from the simula-tion in (a)?

(d) Summarize in a paragraph your findings about the effectiveness of this literature course. Be sure to mention some of the experimental design considerations first raised in Activity 21-11.

Activity 22-5: Wording of Surveys

Much research goes into identifying factors which can unduly influence people's responses to survey questions. In a 1974 study, researchers con-jectured that people are prone to acquiesce, to agree with attitude state-ments presented to them. They investigated their claim by asking subjects whether they agree or disagree with the following statement. Some subjects were presented with form A and others with form B:

- *Form A:* "Individuals are more to blame than social conditions for crime and lawlessness in this country."
- *Form B:* "Social conditions are more to blame than individuals for crime and lawlessness in this country."

The responses are summarized in the table:

	Blame individuals	Blame social conditions
Form A	282	191
Form B	204	268

(a) Let θ_A represent the population proportion of all potential form A subjects who would contend that individuals are more to blame, and let θ_B denote the population proportion of all potential form B subjects who would contend that individuals are more to blame. If the researchers' claim about acquiescence is valid, should θ_A be greater than θ_B—or vice versa?

(a) Calculate the sample proportion of form A subjects who contended that individuals are more to blame. Then calculate the sample proportion of form B subjects who contended that individuals are more to blame.

(c) Conduct (either by hand or with the calculator) the appropriate test of significance to assess the researchers' conjecture. Report the null and alternative hypotheses, the test statistic, and p-value. Also write a one-sentence conclusion.

Activity 22-6: Wording of Surveys *(cont.)*

Researchers have conjectured that the use of the words "forbid" and "allow" can affect people's responses to survey questions. In a 1976 study one group of subjects was asked, "Do you think the United States should forbid public speeches in favor of communism?" while another group was asked, "Do you think the United States should allow public speeches in favor of communism?" Of the 409 subjects asked the "forbid" version, 161 favored the forbidding of communist speeches. Of the 432 subjects asked the "allow" version, 189 favored allowing the speeches.

(a) Calculate the sample proportion of "forbid" subjects who oppose communist speeches (i.e., favor forbidding them) and the sample proportion of "allow" subjects who oppose communist speeches (i.e., do *not* favor allowing them).

(b) Conduct (either by hand or with the calculator) the appropriate two-sided test of significance to test the researchers' conjecture. Report the null and alternative hypotheses (explaining any symbols that you use) as well as the test statistic and p-value. Indicate whether the difference in sample proportions is statistically significant at the .10, .05, and .01 significance levels.

(c) A 1977 study asked 547 people, "Do you think the government should forbid the showing of X-rated movies?" 224 answered in the affirmative (i.e., to forbid). At the same time a group of 576 people was asked, "Do you think the government should allow the showing of X-rated movies?" 309 answered in the affirmative (i.e., to allow). Repeat (a) and (b) for the results of this experiment. (Be very careful to calculate and compare relevant sample proportions. Do not just calculate proportions of "affirmative" responses.)

(d) A 1979 study asked 607 people, "Do you think the government should forbid cigarette advertisements on television?" 307 answered in the affirmative (i.e., to forbid). At the same time a group of 576 people was asked, "Do you think the government should allow cigarette advertisements on television?" 134 answered in the affirmative (i.e., to allow). Repeat (a) and (b) for the results of this experiment; heed the same caution as in (c).

(e) Write a paragraph summarizing your findings about the impact of forbid/allow distinctions on survey questions.

Activity 22-7: Questioning Smoking Policies

An undergraduate researcher at Dickinson College examined the role of social fibbing (the tendency of subjects to give responses that they think the interviewer wants to hear) with the following experiment. Students were asked, "Would you favor a policy to eliminate smoking from all buildings on campus?" For half of the subjects, the interviewer was smoking a cigarette when asking the question; the other half were interviewed by a nonsmoker. Prior to conducting the experiment, the researcher suspected that students interviewed by a smoker would be less inclined to indicate that they favored the ban. It turned out that 43 of the 100 students interviewed by a smoker favored the ban, compared to 79 of the 100 interviewed by a nonsmoker. Carry out the appropriate test of significance to assess the researcher's hypothesis. Write a brief conclusion as if to the researcher, addressing in particular the question of how likely it is that her experimental results would have occurred by chance alone.

WRAP-UP

You have investigated tests of significance for comparing proportions between two experimental groups in this topic. You have used physical (card shuffling) and calculator simulations to study the reasoning behind this procedure; you have also explored the effects of sample sizes on the procedure.

In the next topic you will learn to use confidence intervals to estimate the magnitude of a difference between two proportions. You will also explore the use of these inference procedures with data collected in observational studies.

COMPARING TWO PROPORTIONS II

In the previous topic you discovered that one can use tests of significance to compare proportions between two experimental groups. In this topic you will learn that one can also use confidence intervals to estimate the magnitude of the difference in proportions between the experimental groups. You will also study the application of significance tests and confidence intervals to data obtained from observational studies. While you will find that the reasoning and interpretation of these inference procedures is the same in every situation, you will be reminded that the design of the study dictates the types of conclusions that one can draw.

- To learn a procedure for constructing confidence intervals to estimate differences in proportions between two groups.
- To understand the application of inference procedures for comparing two proportions to data collected from observational studies.
- To continue to recognize and appreciate the important differences between controlled experiments and observational studies with regard to the conclusions one can draw from each.
- To acquire more experience applying inference techniques for comparing two proportions to genuine data.

PRELIMINARIES

1. Take a guess as to the proportion of U.S. college students who drink alcohol.

2. Would you guess that this proportion of U.S. college students who drink alcohol increased, decreased, or remained the same between 1982 and 1991?

3. Of those U.S. college students who drink alcohol, guess the proportion who report getting into a fight after drinking.

4. Would you guess that this proportion of college drinkers who report getting into a fight after drinking alcohol increased, decreased, or remained the same between 1982 and 1991?

5. Of the references to sex on television, take a guess as to the proportion that refer to sex between married partners.

IN-CLASS ACTIVITIES

Recall from the previous topic that you discovered a significance test for the equality of two population proportions:

$$H_0 : \theta_1 = \theta_2$$
$$H_a : \theta_1 < \theta_2, \quad \text{or} \quad H_a : \theta_1 > \theta_2, \quad \text{or} \quad H_a : \theta_1 \neq \theta_2$$

$$\text{Test statistic: } z = \frac{\hat{p}_1 - \hat{p}_2}{\sqrt{\hat{p}_c(1 - \hat{p}_c)\left(\dfrac{1}{n_1} + \dfrac{1}{n_2}\right)}}$$

$$\text{p-value} = \Pr(Z < z) \text{ or } \Pr(Z > z) \text{ or } 2\,\Pr(Z > |z|)$$

One can estimate the magnitude of the difference $\theta_1 - \theta_2$ in the proportions with a confidence interval as follows:

$$(\hat{p}_1 - \hat{p}_2) \pm z^* \sqrt{\frac{\hat{p}_1(1 - \hat{p}_1)}{n_1} + \frac{\hat{p}_2(1 - \hat{p}_2)}{n_2}}$$

where $\hat{p}_1$ and $\hat{p}_2$ are the respective *sample* proportions from the two groups, n_1 and n_2 are the respective sample sizes of the two groups, and z^* is the appropriate critical value from the standard normal distribution.

The formal assumptions that underlie the validity of these procedures are the following:

1. The two samples are independently selected simple random samples from the populations of interest.
2. Both sample sizes are large ($n_1 \geq 30$ and $n_2 \geq 30$ as a rule-of-thumb).

Activity 23-1: Pregnancy, AZT, and HIV (*cont.*)

Consider again your analysis of the AZT experiment from Activity 22-2. Continue to let θ_{AZT} denote the proportion of *all* potential AZT-takers who would have HIV-positive babies and θ_{plac} denote the proportion of *all* potential placebo-takers who would have HIV-positive babies.

(a) Find (by hand) a 95% confidence interval for estimating the difference in population proportions $\theta_{AZT} - \theta_{plac}$.

(b) Write a one-sentence description of what this interval says about the difference in population proportions. In particular, comment on the meaning of whether the interval includes the value zero, contains only negative values, or contains only positive values. (Please be sure that your comments relate to this context of HIV-positive babies.)

(c) Use the 2-PropZInt feature, located in the STAT TESTS menu of your calculator, to find a 98% confidence interval for estimating $\theta_{AZT} - \theta_{plac}$. Record the interval below, and indicate how it differs from the interval you found in (a). Does this interval include zero?

(d) Use the calculator to find a 95% confidence interval for estimating $\theta_{plac} - \theta_{AZT}$. How does this interval differ from the one you found in (a)? How would your conclusion in (b) differ (if at all)?

These inference procedures for comparing two proportions can be used with data from observational studies as well as from controlled experiments. No matter how statistically significant a difference might be, however, one can *not* draw conclusions about *causation* from an observational study.

Activity 23-2: Campus Alcohol Habits

Some researchers wanted to investigate whether the proportion of college students who drink alcohol decreased between 1982 and 1991. They analyzed data from two national studies. In a national study conducted in 1982, 4324 of a sample of 5252 college students said that they drank alcohol. In a similar national study conducted in 1991, 3820 of a sample of 4845 college students said that they drank alcohol.

(a) What proportion of the 1982 sample drank alcohol? What proportion of the 1991 sample drank alcohol?

(b) State the null and alternative hypotheses for testing the researchers' conjecture in both symbols and words.

(c) Use the calculator to compute the test statistic and the p-value of the test. Report these below. Is the decrease in sample proportions statistically significant at the .01 level?

(d) Have the calculator produce a 95% confidence interval for the difference in proportions. Report the interval below. Does the interval include the value 0? What does the interval reveal about the question of whether the proportion of alcohol consumers dropped between 1982 and 1991?

(e) Is this study a controlled experiment or an observational study?

(f) Does the design of the study allow you to supply a causal explanation for the decrease in proportions of alcohol drinkers?

Activity 23-3: Berkeley Graduate Admissions (*cont.*)

Recall from Activity 11-13 that the University of California at Berkeley was alleged to have committed discrimination against women in its graduate admissions practices. When you analyzed the data from the six largest graduate programs, you found that 1195 of the 2681 male applicants had been accepted and that 559 of 1835 female applicants gained acceptance.

(a) Use the calculator to conduct the appropriate test of significance for assessing whether this difference in sample proportions is statistically significant. Report the null and alternative hypotheses, the test statistic, and the p-value of the test. Are these sample results statistically significant at commonly used levels of significance?

(b) Remembering the analysis that you performed in Activity 11-13, do you regard this statistically significant difference as evidence of discrimination? Explain.

HOMEWORK ACTIVITIES

Activity 23-4: College Students' Credit (*cont.*)

Refer back to the data recorded in the "Preliminaries" section of Topic 18 concerning whether college students have their own credit cards.

(a) Calculate the sample proportion of college women who have their own credit card and the sample proportion of college men who have their own credit card.

(b) Conduct (by hand) a test of significance to assess whether these sample proportions differ significantly at the $\alpha = .05$ significance level. Report the null and alternative hypotheses, identifying whatever symbols you introduce in the context of this application. Also show the details of the calculation of the test statistic and p-value.

(c) Estimate the difference in these population proportions with a 95% confidence interval. Also write a sentence explaining what the interval reveals.

(d) Comment on the connection between the result of the significance test in (b) and the confidence interval in (c).

(e) Is this study a controlled experiment or an observational study? Explain your answer.

Activity 23-5: Age and Political Ideology *(cont.)*

Reconsider the data tabulated in Activity 11-2 concerning age and political ideology. Of the 296 people in the "under 30" age group, 83 identified themselves as politically liberal. Of the 586 people in the "over 50" age group, 88 regarded themselves as liberal.

(a) Conduct (by hand or with the calculator) a test of significance to determine if these sample results provide strong evidence that the proportion of liberals among all "under 30" people differs from that among all "over 50" people.

(b) Estimate the difference in these population proportions with a confidence interval. (Choose your own confidence level.)

(c) Write a few sentences describing your findings from (a) and (b) regarding the question of whether the proportion of liberals differs between "under 30" and "over 50" people and, if so, by about how much.

Activity 23-6: BAP Study

Researchers investigating the disease Bacillary Angiomatosis and Peliosis (BAP) conducted a study of 48 BAP patients and a control group of 94 subjects. The following table lists the numbers of people in each group who had the indicated characteristics.

	Case patients (n=48)	Control group (n=94)
# Male	42	84
# White	43	89
# Non-Hispanic	38	75
# With AIDS	24	44
# Who own a cat	32	37
# Scratched by a cat	30	29
# Bitten by cat	21	14

(a) Is this study a controlled experiment or an observational study? If it is an observational study, what kind is it? Explain.

(b) Use the calculator to perform tests of significance on each of the variables listed in the table. On which variables do the two groups differ significantly at the .05 level?

(c) In light of the type of study involved, can you conclude that these variables cause the disease? Explain.

Activity 23-7: Baldness and Heart Disease *(cont.)*

Reconsider the data presented in Activity 11-14 and mentioned in Activity 20-13 concerning baldness and heart disease.

(a) Calculate the sample proportion of the heart disease patients who had "some" baldness or more (i.e., some, much, or extreme baldness). Calculate the same for the control group.

(b) Conduct (by hand or with the calculator) a test of significance to determine if these proportions differ significantly. State the null and alternative hypotheses in words as well as in symbols, and report the test statistic and p-value. Also indicate whether the difference in sample proportions is statistically significant at the .05 level.

(c) Construct (by hand or with the calculator) a 97.5% confidence interval for the difference in population proportions. Write a sentence or two describing what the interval reveals.

(d) What conclusion can you draw about a possible association between baldness and heart disease? Be sure to keep in mind the design of the study as you address this question.

Activity 23-8: Sex on Television

In his book *Hollywood v. America: Popular Culture and the War on Traditional Values*, Michael Medved cites a pair of studies that examined instances of, and references to, sexual activity on prime-time television programs. A 1981 study examined a sample of 47 references to sex on prime-time television shows and found that 6 of the references were to sexual intercourse between partners who were married to each other. A similar (but larger) study in 1991 examined a sample of 615 references to sex and found that 44 of the references were to sex between married partners.

(a) What proportion of the sexual references in the 1981 study were to sex between married partners? In the 1991 study?

(b) Use the calculator to conduct the appropriate significance test of whether the proportion of *all* sexual references which describe married sex has decreased from 1981 to 1991. State the null and alternative hypotheses (identify which group is which), and report the test statistic and p-value of the test. Would you reject the null hypothesis at the $\alpha = .01$ level?

(c) Use the calculator to find a 99% confidence interval for the difference in the proportions of references to married sex among *all* such television references between 1981 and 1991; record the interval below.

(d) Let θ denote the proportion of *all* sexual references on television in 1991 which pertained to sex between married partners. Use the sample data (from the 1991 study only) to find (by hand) a 95% confidence interval for θ. (You may want to refer back to the inference procedures for a population proportion that were developed in Topics 15–17.)

Activity 23-9: Heart By-Pass Surgery

The November 20, 1992, issue of *The Harrisburg Evening-News* reported the failure rates for coronary artery bypass surgery in Pennsylvania in 1990. The newspaper reported the number of patients who underwent the operation and the number who died during or after surgery, broken down by geographic region, by hospital, and by physician. Central Pennsylvania had 125 deaths in 3676 operations, compared to 288 deaths in 6313 operations for the southeastern region and 167 deaths in 4906 operations for the western part of the state.

(a) For each of the three regions (central, southeastern, and western), calculate the proportion of operations that resulted in death. Which region seems to do the best? the worst?

(b) Use the calculator to conduct tests of significance to assess whether the sample proportions of recovery differ significantly between each of the three *pairs* of regions. Record the p-values of the tests, indicate which (if any) of the differences are statistically significant at the .10 level, and write a few sentences summarizing your findings.

(c) What other information would you want to know before you compare physicians and hospitals solely on failure rates of operations? (You might want to look back at the hypothetical data in Activity 11-4 for one idea.)

Activity 23-10: Employment Discrimination

In the legal case of *Teal v. Connecticut* (1982), a company was charged with discrimination in that blacks passed its employment eligibility exam in smaller proportions than did whites. Of the 48 black applicants to take the test during the year in question, 26 passed; of the 259 white applicants to take the test, 206 passed.

(a) What is the sample proportion of black applicants who passed the test? What is the sample proportion of white applicants who passed?

(b) Perform (by hand; you may use the calculator to check your work) the significance test of whether the data provide strong evidence that the proportion of blacks who pass the test is statistically significantly less than that of whites. Show your work, indicate whether the difference is significant at the .05 and/or .01 levels, and write a one- or two-sentence conclusion (as if reporting to the jurors in the case).

Activity 23-11: Campus Alcohol Habits (*cont.*)

Recall the study described in Activity 23-2 concerning alcohol use on campus. The national studies also asked students more specific questions about students' drinking habits. Students were asked whether they had gotten into fights after drinking and whether they have had trouble with the law after drinking. Of the 4324 drinkers surveyed in 1982, 502 reported getting into a fight after drinking, while 190 reported getting into trouble with the law due to drinking. Of the 3820 drinkers surveyed in 1991, these numbers were 657 and 290, respectively.

Use the calculator to analyze these data for any differences between 1982 and 1991 with regard to getting into fights or trouble with the law due to drinking. Report the results of significance tests as well as confidence intervals. Also compare your findings with those from Activity 23-2. Write a paragraph or two summarizing your conclusions.

Activity 23-12: Kids' Smoking

A newspaper account of a medical study claimed that the daughters of women who smoked during pregnancy are more likely to smoke themselves. The study surveyed children, asking them if they had smoked in the last year and then asking the mother if she had smoked during pregnancy. Only 4% of the daughters of mothers who did not smoke during pregnancy had smoked in the past year, compared to 26% of girls whose mothers had smoked during pregnancy.

(a) What further information do you need to determine if this difference in sample proportions is statistically significant?

(b) Suppose that there had been 50 girls in each group. Use the calculator to conduct a two-sided significance test. Report the p-value and whether the difference in sample proportions is statistically significant at the .05 level.

(c) Repeat (b) supposing that there had been 50 girls whose mothers had smoked and 200 whose mothers had not.

(d) Repeat (b) supposing that there had been 200 girls in each group.

(e) Is this study a controlled experiment or an observational study?

(f) Even if the difference in sample proportions is statistically significant, does the study establish that the pregnant mother's smoking caused the daughter's tendency to smoke? Explain.

Activity 23-13: Kids' Smoking (*cont.*)

Refer back to the description of the study concerning kids' smoking in Activity 22-12. The researchers also studied sons and found that 15% of the sons of mothers who had not smoked during pregnancy had smoked in the past year, compared to 20% of the sons of mothers who had smoked.

(a) Suppose that there had been 60 boys in each group. Use the calculator to conduct a two-sided significance test. Report the p-value and whether the difference in sample proportions is statistically significant at the .05 level.

(b) Repeat (a) supposing that there had been 200 boys in each group.

(c) Repeat (a) supposing that there had been 500 boys in each group.

(d) Suppose that the two groups had the same number of boys. Try to find the smallest number for this sample size which would make the difference in sample proportions statistically significant at the .05 level. You may either use trial-and-error with the calculator or solve the problem analytically by hand.

Activity 23-14: Hypothetical Medical Recovery Rates (*cont.*)

Recall the situation of Activity 22-3 in which a new medical treatment is compared to an old, standard treatment. Suppose that very large samples of 50,000 are used for each treatment and that the sample proportions of recovery are 86.7% for the old, and 87.3% for the new treatment.

(a) Use the calculator to conduct a test of significance for assessing whether the new treatment has a statistically significantly higher recovery rate than the old treatment. Report the null and alternative hypotheses, the test statistic, and the p-value. Is the result statistically significant at the .01 level?

(b) Use the calculator to construct a 99% confidence interval for estimating the difference in population proportions of recovery. Record the interval and comment on whether it contains only positive values, only negative values, or some of each.

(c) Recall the distinction between statistical significance and practical significance introduced in Activity 20-4. Explain how and why this distinction pertains to the comparison of these medical treatments.

Activity 23-15: Comparing Proportions of Personal Interest

Write a paragraph detailing a real situation in which you would be interested in performing a test of significance to compare two *proportions*. Describe precisely the context involved and explain whether the study would be a controlled experiment or an observational study. Also suggest how you might go about collecting the sample data.

WRAP-UP

This topic has broadened your knowledge of inference procedures for comparing two proportions by introducing you to confidence intervals to accompany tests of significance. You have also examined the use of these inference procedures with data from observation studies.

To this point in the text you have applied inference procedures only to data from *categorical binary* variables; in other words, you have dealt only with *proportions*. In the next unit you will expand your knowledge of inference procedures by considering *measurement* variables and learning procedures for making inferences about population *means*.

Unit Six

Inference from Data: Measurements

Topic 24:

INFERENCE FOR A POPULATION MEAN I

OVERVIEW

You have explored fundamental principles of statistical inference while applying them to problems involving a population *proportion* or a comparison of two population *proportions*. In other words, all of the inference procedures you have studied pertain to *binary categorical* variables. Throughout the remainder of the text you will study the application of inference techniques to *measurement* variables.

In this topic you will begin to study the principles involved in inferences about the *mean* of a population. You will encounter and explore one of the most famous of statistical techniques: the *t-test*.

OBJECTIVES

- To discover which sample statistics play roles in making inferences about a population mean.
- To become familiar with the use of a table of the *t-distribution* for finding critical values and p-values.
- To learn specific details of producing confidence intervals and conducting significance tests for a population mean.
- To investigate the role of factors such as sample size and sample variability with regard to confidence intervals and significance tests for a population mean.
- To recognize both legitimate and inappropriate interpretations of inference procedures for measurement variables.

PRELIMINARIES

1. Take a guess as to the least amount of sleep that a student in your class got last night. Also make a guess for the most sleep enjoyed by a student in your class last night.

2. Mark on the scale below an interval that you believe with 90% confidence to include the mean amount of sleep (in hours) that a student at your school got last night.

```
---+---------+---------+---------+---------+---------+---.
   5.0       6.0       7.0       8.0       9.0       10.0
```

3. Mark on the scale below an interval that you believe with 99% confidence to include the mean amount of sleep (in hours) that a student at your school got last night.

```
---+---------+---------+---------+---------+---------+---.
   5.0       6.0       7.0       8.0       9.0       10.0
```

4. Which of these two intervals is wider?

5. Record in the table below the bed time and wake time of the students in your class last night. Also calculate and record the sleeping times; be careful to measure sleeping times in minutes.

Student	Bedtime	Wake time	Sleep (min.)	Student	Bedtime	Wake time	Sleep (min.)
1				13			
2				14			
3				15			
4				16			
5				17			
6				18			
7				19			
8				20			
9				21			
10				22			
11				23			
12				24			

IN-CLASS ACTIVITIES

As you have discovered in previous units, the distinctions between *popula-tion* and *sample* and between *parameter* and *statistic* are at the heart of statistical inference. The whole point of inference is to learn something about unknown population parameters on the basis of observed sample statistics.

Recall that when dealing with measurement variables, the mean and standard deviation of the *population* are typically denoted by μ and σ, respectively; the mean and standard deviation of the *sample* are typically denoted by $\bar{x}$ and s, respectively.

Activity 24-1: Parameters vs. Statistics (*cont.*)

Identify each of the following as a parameter or a statistic; also indicate the symbol that would typically be used to denote the value.

(a) the mean time spent sleeping last night by the students in your class

(b) the mean time spent sleeping last night by *all* students at your school

(c) the standard deviation of the sleeping times for the students in your class

(d) the standard deviation of the sleeping times for *all* students at your school

(e) the mean number of runs scored in all Major League baseball games played in 1992

(f) the mean number of runs scored in a simple random sample of 30 baseball games from the 1992 Major League season

(g) the mean number of states visited among all students at your school

(h) the standard deviation of the distances from home for the students in your class

(i) the mean age of all pennies currently in circulation in the United States

(j) the mean number of keys carried by a student at your school

(k) the mean IQ of subjects in a particular study who claim to have had experiences with UFO's.

Activity 24-2: Students' Sleeping Times

Suppose that you want to estimate the mean sleeping time of *all* students at your school last night. Suppose that you also want to assess whether this mean (which we denote by μ) is less than eight hours.

(a) Hypothetical sleep times have been stored in a grouped file named HYPOSLEEP.83g. Download this grouped file into your calculator.

(b) Consider the four different (hypothetical) samples of sleeping times presented in the dotplots below. Use the calculator to compute the sample size, sample mean, and sample standard deviation for each sample, recording the results in the table below. (The data appear in Appendix A.)

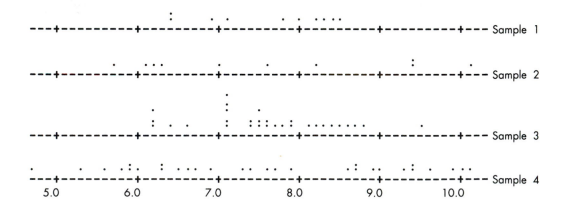

Hypothetical sample	Sample size	Sample mean	Sample std. dev.
1			
2			
3			
4			

(c) What do all of these samples have in common?

(d) What strikes you as the most important difference between the distribution of sleeping times in Sample 1 and in Sample 2?

(e) Which of these two samples (1 or 2) do you think would produce a more accurate estimate of μ; i.e., which sample would produce a narrower confidence interval for μ?

(f) Which of these two samples (1 or 2) do you think supplies stronger evidence that $\mu < 8$; i.e., which sample would produce a smaller p-value of the appropriate test of significance?

(g) What strikes you as the most important difference between the distribution of sleeping times in Sample 1 and in Sample 3?

(h) Which of these two samples (1 or 3) do you think would produce a more accurate estimate of μ; i.e., which sample would produce a narrower confidence interval for μ?

(i) Which of these two samples (1 or 3) do you think supplies stronger evidence that $\mu < 8$; i.e., which sample would produce a smaller p-value of the appropriate test of significance?

(j) With which of the *four* samples do you think you would be able to estimate μ *most* accurately; i.e., which sample do you think would produce the *narrowest* confidence interval for μ?

(k) With which of the *four* samples do you think you would be able to estimate μ *least* accurately; i.e., which sample do you think would produce the *widest* confidence interval for μ?

(l) What information from a simple random sample do you suspect one needs to know in order to make inferences about the population mean?

We hope that your intuition has suggested that in addition to knowing the sample mean, one needs to consider the sample size and sample variability (as measured by the sample standard deviation) when making inferences about a population mean.

The form of a confidence interval for a population mean is very similar to that for a population proportion (which you have already studied in depth), and the interpretation of confidence intervals is always the same. The idea is to start with a sample statistic (in this case, the *sample mean $\bar{x}$*) which estimates the parameter of interest (in this case, the *population mean μ*). One then goes a certain distance to either side of that sample statistic;

this distance is found by multiplying a critical value (which depends on the confidence level desired) by an estimate of the standard deviation of the sample statistic. In this case, the critical value comes not from the standard normal distribution but from a *t-distribution* (which you will soon investigate), and the estimate of the standard deviation of the sample statistic is $\frac{s}{\sqrt{n}}$, where s denotes the standard deviation of the sample values and n the sample size.

Thus, the form for a confidence interval for a population mean μ is given by

$$\bar{x} \pm t^{*}_{n-1} \frac{s}{\sqrt{n}},$$

where t^{*}_{n-1} is the appropriate critical value from the *t*-distribution with $n - 1$ *degrees of freedom*.

This *t*-distribution is actually an entire family of distribution curves, not unlike the normal distributions. Each member of the family is characterized by an integer number called its *degrees of freedom* (abbreviated d.f.). These *t*-distributions have the following properties:

- They are symmetric about 0 (as with the standard normal distribution).
- They are mound-shaped (as with the standard normal distribution).
- They are more spread out than the standard normal distribution; i.e., they have wider, fatter tails than the standard normal distribution.
- The larger the degrees of freedom, the narrower the tails; in fact, as the degrees of freedom get infinitely large, the *t*-distribution approaches the standard normal distribution.

Just as you have used a standard normal distribution table, you will now see how to use a *t*-table such as Table II at the back of the book. Notice that each line of the table corresponds to a different value of the degrees of freedom. Always start by going to the line that is relevant for the degrees of freedom with which you are working. Next note that the table gives various values of p across the top of the table. Finally, observe that the table gives values such that the probability of lying to the *right* of that value (equivalent to the area to the right of that value under the *t*-distribution) is p.

For instance, if you are dealing with a *t*-distribution with 5 degrees of freedom, the table tells you that the probability of lying to the right of 2.015 is .05. (Please verify this for yourself.) We will convert this statement into symbols by writing $Pr(T_5 > 2.015) = .05$.

Activity 24-3: Exploring the *t*-Distribution

Find the critical value t^* for a 95% confidence interval based on 11 degrees of freedom by following these steps:

(a) Draw a rough sketch of the *t*-distribution with 11 d.f.:

(b) The critical value t^* for a 95% confidence interval is the value such that 95% of the area under the curve is between $-t^*$ and t^*. Shade this area on your sketch above.

(c) What is the area to the *right* of t^* under the curve?

(d) Look in the *t*-table to find the value t^* which has that area to its right under a *t*-distribution with 11 d.f.

(e) Is this critical value less than or greater than the critical value z^* from the standard normal distribution for a 95% confidence interval?

Critical values t^* from the *t*-distribution are always greater than their counterparts from the *z*- (standard normal) distribution.

(f) Use the *t*-table to find the critical value t^* for a 95% c.i. based on 23 d.f.

(g) Use the *t*-table to find the critical value t^* for a 99% c.i. based on 40 d.f.

(h) Find the critical value for a 95% c.i. based on 94 d.f. (When the exact degrees of freedom do not appear in the table, always round the degrees of freedom *down* to be conservative.)

The structure of significance tests concerning a population mean is also very similar to that for a population proportion, and the reasoning and inter-pretation of such tests is the same in all circumstances. The chief differences are that the sample mean, sample standard deviation, and sample size all enter into the calculation of the test statistic, and the p-value is found by comparing the test statistic to a *t*-distribution rather than to the standard nor-mal distribution.

The form of a test of significance concerning a population mean μ is given by:

$$H_0 : \mu = \mu_0$$

$$H_a : \mu < \mu_0, \quad \text{or} \quad H_a : \mu > \mu_0, \quad \text{or} \quad H_a : \mu \neq \mu_0$$

$$\text{Test statistic: } t = \frac{\bar{x} - \mu_0}{s/\sqrt{n}}$$

p-value: $\Pr(T_{n-1} < t)$, or $\Pr(T_{n-1} > t)$, or $2\,\Pr(T_{n-1} > |t|)$

where μ_0 represents the hypothesized value of the population mean and T_{n-1} represents a *t*-distribution with $n - 1$ degrees of freedom.

Thus, one uses the *t*-table to find the p-value of the significance test. For example (just as above), if the alternative hypothesis were $H_a : \mu > \mu_0$, the degrees of freedom were 5, and the test statistic turned out to be $t = 2.015$, then the p-value of the test would be .05, since $\Pr(T_5 > 2.015) = .05$. If the test statistic, though, were 2.206, then the limitations of the *t*-table would only allow one to say that the p-value falls between .025 and .05. (Of course, the calculator could calculate the p-value much more exactly.) Typically, the *t*-table only reveals a range of values in which the p-value must fall.

Activity 24-4: Exploring the *t*-Distribution (*cont.*)

(a) Use the *t*-table to find $\Pr(T_7 > 1.415)$.

(b) Use the *t*-table to find $\Pr(T_7 > 2.517)$.

(c) Use the *t*-table to find (as accurately as you can) $\Pr(T_7 > 2.736)$.

(d) Use the t-table to find (as accurately as you can) $\Pr(T_7 < -2.736)$. (Be clever here: draw a sketch of the t-distribution, shade in the area you want, and use the symmetry of the t-distribution to determine this area.)

(e) Use the t-table to find (as accurately as you can) $\Pr(T_{24} > 3.168)$.

(f) Use the t-table to find (as accurately as you can) $\Pr(T_{24} > 3.926)$.

(g) Use the t-table to find (as accurately as you can) $\Pr(T_{24} > 0.628)$.

(h) Use the t-table to find (as accurately as you can) $2 \Pr(T_{24} > 3.168)$.

The two formal assumptions which underlie the validity of these inference techniques bear mentioning. The first (as always) is that the sample is a simple random sample from the population of interest. The second assumption is *either* that the sample size is large ($n \geq 30$ as a rule-of-thumb) *or* that the population is normally distributed. (Notice that only one, not both, of these assumptions needs to be satisfied in order for the inference procedure to be valid.)

Finally, be aware that the reasoning behind and the interpretation of these inference procedures are the same as always. For instance, a 95% confidence interval indicates that one is 95% confident that the interval contains the true value of the population mean; if one were to repeat the procedure over and over, in the long run 95% of the intervals so generated would contain the population mean. Similarly, the p-value of the significance test reveals the probability, assuming the null hypothesis to be true, of having obtained a sample result at least as extreme as the one actually attained; thus, the *smaller* the p-value, the stronger the evidence *against* the null hypothesis.

Activity 24-5: Students' Sleeping Times (*cont.*)

(a) Reconsider the four hypothetical samples of students' sleeping times presented in Activity 24-2. Re-record the sample sizes, sample means, and sample standard deviations in the table below.

(b) Use the TInterval feature (found in the STAT TESTS menu of the calculator) to produce a 95% confidence interval for μ (the mean sleeping time for all students of the school on the night in question), recording the interval in the table.

(c) Use the T-Test feature (found in the STAT TESTS menu of the calculator) to conduct a significance test of whether the sample provides evidence that μ is actually less than eight hours. Record the p-value of the test in the table.

Hypothetical sample	Sample size	Sample mean	Sample std. dev.	95% confidence interval	(One-sided) p-value
1					
2					
3					
4					

(d) Based on these results, indicate which hypothetical sample estimates μ *most* accurately (i.e., with the narrowest confidence interval), which estimates μ *least* accurately (i.e., with the widest confidence interval), which provides the *most* evidence that $\mu < 8$ (i.e., has the smallest p-value), and which provides the *least* evidence that $\mu < 8$ (i.e., has the largest p-value):

Most accurate:

Least accurate:

Most evidence:

Least evidence:

Activity 24-6: Students' Sleeping Times (*cont.*)

Now consider the data on sleeping times collected from the students in your class.

(a) Enter the data into the calculator, and convert the unit of measurement to hours (by dividing the minutes by 60). Then use the calculator to produce visual displays of the distribution. Write a few sentences commenting on key features of this distribution.

(b) Use the calculator to compute the sample size, sample mean, and sample standard deviation. Record these values below, and indicate the symbol used to represent each.

(c) Let μ denote the mean sleeping time for *all* students at your school on that particular night. Perform (by hand) the appropriate significance test to address whether the sample data provide much evidence that μ is less than eight hours. (You may use the calculator to check your work.) Also write a one-sentence conclusion (based on the p-value of the test).

(d) Find (by hand) a 95% confidence interval for μ. (You may again use the calculator to check your work.)

(e) If the sample size had been larger (and all else had turned out the same), how would the confidence interval and the p-value of the test have been affected?

(f) If the sample standard deviation had been larger (and all else had turned out the same), how would the confidence interval and the p-value of the test have been affected?

(g) If the sample mean had been smaller (and all else had turned out the same), how would the confidence interval and the p-value of the test have been affected?

(h) Comment on whether the assumptions that underlie these inference procedures seem to be satisfied in this case.

(i) If this were a class which meets at 8:00 am, would you consider the inference results valid for the population of all students at your school? Explain.

HOMEWORK ACTIVITIES

Activity 24-7: Exploring the *t*-Distribution (*cont.*)

(a) Use the *t*-table to find the critical values t^* corresponding to the following confidence levels and degrees of freedom, filling in a table as below with those critical values:

D.F.\Conf. level	90%	95%	99%
4			
11			
23			
40			
80			
Infinity			

(b) Does the critical value t^* get larger or smaller as the confidence level gets larger (if the degrees of freedom remain the same)?
(c) Does the critical value t^* get larger or smaller as the degrees of freedom get larger (if the confidence level remains the same)?
(d) Do the critical values from the *t*-distribution corresponding to infinite degrees of freedom look familiar? Explain. (Refer back to Topic 16 if they do not.)

Activity 24-8: Exploring the *t*-Distribution (*cont.*)

(a) Use the *t*-table to find the p-values (as accurately as possible) corresponding to the following test statistic values and degrees of freedom, filling in a table like the following with those p-values:

D.F.	$Pr(T > 1.415)$	$Pr(T > 1.960)$	$Pr(T > 2.517)$	$Pr(T > 3.168)$
4				
11				
23				
40				
80				
Infinity				

(b) Does the p-value get larger or smaller as the value of the test statistic gets larger (if the degrees of freedom remain the same)?

(c) Does the p-value get larger or smaller as the degrees of freedom get larger (if the value of the test statistic remains the same)?

Activity 24-9: Students' Sleeping Times (*cont.*)

Consider your analysis of students' sleeping times from Activity 24-6. Identify each of the following statements as legitimate or illegitimate interpretations of your results. (You may want to refer back to the interpretations of confidence intervals and p-values given in Unit IV.)

- One can be 95% confident that the interval contains the true value of μ.
- If one repeatedly took random samples of college students and generated 95% confidence intervals in this manner, then in the long run 95% of the intervals so generated would contain the true value of μ.
- The probability is .95 that μ lies in the interval.
- One can be 95% confident that the sleeping time for any particular student falls within the interval.
- 95% of the students in the sample had sleeping times that fall within the interval. (Check this for yourself by counting how many of the students in the sample had sleeping times which fall within this interval.)
- The p-value is the probability that $\mu < 8$.
- The p-value is the probability of the sample mean having been as small or smaller than it actually was if μ were equal to 8.

Activity 24-10: Students' Sleeping Times (*cont.*)

Reconsider the hypothetical samples of sleeping times presented in Activity 24-2 and analyzed in Activity 24-5 above. Suppose now that you were interested in testing whether the sample data provide evidence that the population mean sleeping time *differs from* eight hours. Use the calculator to conduct the test of significance for each of the four samples with a *two-sided* alternative hypothesis. Record the p-value for each sample and comment on how these p-values compare with the one-sided ones found in Activity 24-5.

Activity 24-11: UFO Sighters' Personalities (*cont.*)

Consider the study described in Activity 21-7 dealing with individuals who claim to have had experiences with UFO's. The sample mean IQ of the 25 people in the study who claimed to have had an intense experience with a UFO was 101.6; the standard deviation of these IQ's was 8.9.

(a) Use this sample information to produce (by hand) a 95% confidence interval for the population mean IQ of all people who have had intense UFO experiences.

(b) Conduct (by hand) a test of significance for assessing whether the sample data support the belief that the mean IQ among all people who have had intense UFO experiences exceeds 100. Write a one-sentence conclusion as well as reporting the details of the test.

WRAP-UP

This topic has introduced you to inference procedures for a population mean. You have found that the reasoning behind and interpretation of confidence intervals and significance tests is exactly the same as you have already studied; the details of the calculations are all that change as you apply these procedures in different settings.

You have discovered that the sample size and sample standard deviation play a role in making inferences about a population mean, as does the sample mean. You have encountered the *t-distribution* as an important component of these inference procedures and garnered a great deal of practice using a *t*-table.

In the next topic you will apply these inference procedures to a variety of problems involving genuine data. You will also discover a special type of experimental design which utilizes this procedure—the *paired comparisons* design.

Topic 25:

INFERENCE FOR A POPULATION MEAN II

OVERVIEW

You have studied in a preliminary fashion the application of confidence intervals and significance tests to measurement variables for which the population mean is the parameter of interest. You will apply these procedures to a variety of genuine applications in this topic. You will also discover the *paired comparisons design* as a convenient and useful method of comparing two treatments, and you will see how to use these inference techniques to analyze data collected with such a design.

OBJECTIVES

- To acquire experience at applying inference techniques to a variety of practical problems involving population means.
- To continue to develop skills at interpreting the results of the inference procedures.
- To become re-acquainted with the skills of analyzing distributions graphically, numerically, and verbally.
- To understand the idea of the *paired comparisons design* as a useful experimental design for comparing two treatments.
- To develop the ability to analyze data from a paired comparisons experiment.

PRELIMINARIES

1. Take a guess as to the mean number of keys carried by a student at your school.

2. Mark on the scale below an interval that you believe with 95% confidence to contain the population mean number of keys carried by students at your school.

```
+---------+---------+---------+---------+---------+-
0.0       2.0       4.0       6.0       8.0       10.0
```

3. Record in the table below the number of keys carried by each student in your class today:

Student	# Keys	Student	# Keys	Student	# Keys	Student	# Keys
1		7		13		19	
2		8		14		20	
3		9		15		21	
4		10		16		22	
5		11		17		23	
6		12		18		24	

4. Mark on the scale below an interval that you believe with 95% confidence to contain the population mean number of states visited by students at your school.

```
+---------+---------+---------+---------+---------+-
0         10        20        30        40        50
```

5. Mark on the scale below an interval that you believe with 95% confidence to contain the population mean difference in marriage ages (husband's age minus wife's age).

```
+---------+---------+---------+---------+---------+-
-6.0      -3.0      0.0       3.0       6.0       9.0
```

IN-CLASS ACTIVITIES

Remember the forms of confidence intervals and significance tests concerning a population mean μ that you studied in the previous topic (where n continues to denote the sample size, $\bar{x}$ the sample mean, and s the sample standard deviation):

   Confidence interval for a population mean μ:

$$\bar{x} \pm t^*_{n-1} \frac{s}{\sqrt{n}}$$

where t^*_{n-1} is the appropriate critical value from the t-distribution with $n-1$ degrees of freedom.

Significance test about a population mean μ:

$$H_0 : \mu = \mu_0$$
$$H_a : \mu < \mu_0, \quad \text{or} \quad H_a : \mu > \mu_0, \quad \text{or} \quad H_a : \mu \neq \mu_0$$

Test statistic: $t = \dfrac{\bar{x} - \mu_0}{s/\sqrt{n}}$

p-value: $\Pr(T_{n-1} < t)$, or $\Pr(T_{n-1} > t)$, or $2\,\Pr(T_{n-1} > |t|)$

---

where μ_0 represents the hypothesized value of the population mean and T_{n-1} represents a t-distribution with $n-1$ degrees of freedom.

Assumptions:

1. The sample is a simple random sample (SRS) from the population of interest.
2. The sample size is large ($n \geq 30$ as a rule-of-thumb) *or* that the population is normally distributed.

Remember also how to use the t-table to determine critical values and to compute p-values. Finally, recall that the reasoning behind and interpretation of these inference procedures is the same as you studied earlier; in fact, these do not change from situation to situation.

Activity 25-1: Students' Travels (*cont.*)

Refer back to the data on students' travels that were collected in Topic 2.

(a) Enter the number of *states* visited by each student into the calculator and compute the sample size, sample mean, and sample standard deviation. Record these below along with the symbols used to represent them.

(b) Find (by hand) a 90% confidence interval for μ, the population mean number of states visited, among *all* students at the college.

(c) Look back at the original data to determine how many of the *sample* values fall within this confidence interval. In other words, for how many students does the number of states visited fall in this interval?

(d) What proportion of the sample is this? Should this proportion be close to 90%? Explain.

Confidence intervals of this type estimate the value of a population *mean*. They do not estimate the values of *individual* observations in the population or in the sample.

(e) If the sample size had been *four times* its actual size and if the sample mean and standard deviation had been the same, how would the half-width of the 90% confidence interval for μ change? (Be specific about your answer.)

(f) If the sample size had been *nine times* its actual size and if the sample mean and standard deviation had been the same, how would the half-width of the 90% confidence interval for μ change? (Again be specific about your answer.)

(g) Is this a simple *random* sample from the population of all students at the college? If not, do you have any reason to suspect that the sample is not *representative* of the population in terms of states visited? Explain.

One should recognize that convenience samples are never a substitute for random samples. Since convenience samples abound, however, one should also question whether they seem to be *biased* with regard to the variable of interest.

(h) Use the calculator to look at visual displays of the distribution of states visited in the sample. Does this sample distribution provide any reason to doubt that the population of states visited follows a normal (symmetric, mound-shaped) distribution? Explain.

One can investigate the assumption of a normally distributed population by looking for marked departures from normality in the distribution of the sample.

Activity 25-2: Hypothetical ATM Withdrawals (*cont.*)

Refer to the hypothetical ATM withdrawals described in Activity 4-18. These pertain to a sample of withdrawal amounts from three different automatic teller machines.

(a) Use the calculator to compute the sample size, sample mean, and sample standard deviation of the withdrawal amounts for each machine. Also have the calculator determine a 95% confidence interval for the mean withdrawal amount among all withdrawals for each machine. Record the results in the table below.

	Sample size	Sample mean	Sample std. dev.	95% confidence interval
Machine 1				
Machine 2				
Machine 3				

(b) Use the calculator to look at visual displays of the three distributions of withdrawal amounts. Do the distributions look the same or even similar? Write a paragraph comparing and contrasting the three distributions of ATM withdrawals.

This activity should remind you that a mean summarizes just one aspect of a distribution. While the mean is often very important and the focus of most inference procedures, by no means does it completely describe a distribution.

Activity 25-3: Marriage Ages (*cont.*)

Reconsider the data presented in Activity 3-7 concerning the ages at marriage for a sample of 24 couples who obtained their marriage licenses in Cumberland County, Pennsylvania, in 1993. The data are recorded below:

Couple #	Husband	Wife	Difference (husband−wife)	Couple #	Husband	Wife	Difference (husband−wife)
1	25	22		13	25	24	
2	25	32		14	23	22	
3	51	50		15	19	16	
4	25	25		16	71	73	
5	38	33		17	26	27	
6	30	27		18	31	36	
7	60	45		19	26	24	
8	54	47		20	62	60	
9	31	30		21	29	26	
10	54	44		22	31	23	
11	23	23		23	29	28	
12	34	39		24	35	36	

(a) For each couple in the sample, subtract the wife's age from the husband's age. Record the results in the table above.

(b) Create (by hand) a dotplot of the differences in ages for these 24 couples.

(c) Use the calculator to compute the relevant summary statistics regarding the sample of *age differences;* record them below along with the symbols used to represent them.

(d) Use the calculator to conduct a significance test of whether the *mean* of the *population* of *age differences* exceeds zero, suggesting that the husband tends to be older than the wife. State the null and alternative hypotheses, and record the test statistic and p-value. Also indicate whether the sample data are statistically significant at the .05 level.

(e) Use the calculator to find a 90% confidence interval for the mean of the *population* of *age differences*. Comment on whether the interval includes zero.

(f) Write a one- or two-sentence conclusion about the difference in ages between husbands and wives based on your analysis of these data.

This example illustrates a *paired comparisons* experimental design. One can *control* some of the variation in marriage ages by considering *couples* rather than *individuals*. A less sensible design for this study would have been to obtain the ages of one sample of 24 brides and of a separate sample of 24 husbands.

One analyzes paired comparisons data by analyzing *differences* using the inference procedures that you have studied concerning a population mean. In this case, the mean of the population of *age differences* is what one makes inferences about.

Activity 25-4: Planetary Measurements (*cont.*)

Reconsider the data presented in Activity 4-6 which listed (among other things) the distance from the sun for each of the nine planets in our solar system. The mean of the distances turns out to be 1102 million miles, and the standard deviation is 1341 million miles.

(a) Use these statistics to construct (by hand) a 95% confidence interval.

(b) Does this interval make any sense at all? If so, what parameter does it estimate? Explain.

This activity should remind you that statistical inference involves using sample statistics to infer something about population parameters. In this example the nine planets in the solar system comprise the entire population, so it makes no sense to estimate the population mean with a confidence interval. In fact, you *know* the population mean distance of a planet from the sun to be 1102 million miles.

HOMEWORK ACTIVITIES

Activity 25-5: Students' Measurements (*cont.*)

Reconsider the data collected in Topic 3 concerning the heights and armspans of students in your class.

(a) Use the calculator to compute each student's *ratio* of height to armspan (i.e., height divided by wingspan). Then have it calculate the sample size, sample mean, and sample standard deviation of these ratios.
(b) Let μ denote the mean height/armspan ratio among *all* college students. Suppose that you want to test the null hypothesis that μ equals

1 versus the alternative hypothesis that μ differs from 1. Calculate (by hand) the appropriate test statistic.

(c) Use the t-table to find the approximate p-value of the test.

(d) Write a sentence describing precisely what the p-value means in this context.

(e) At the $\alpha = .10$ significance level, do the sample data provide enough evidence to reject the null hypothesis that the mean height/wingspan ratio among *all* college students equals 1?

(f) Find (by hand) a 90% confidence interval for μ, the mean height/armspan ratio among *all* college students. (You may use the calculator to check your work.)

(g) Does this interval include the value 1? Comment on how this question relates to the test of significance that you performed above.

Activity 25-6: Students' Keys

Consider the data on the number of keys carried by students in your class which you gathered in the "Preliminaries" section. Enter the data into the calculator.

(a) Have the calculator compute the five-number summary of the number of keys carried. Use this summary to construct (by hand) a boxplot of the distribution of number of keys carried. Does the sample distribution reveal any marked departures from normality? Explain.

(b) Use the calculator to compute a 96% confidence interval for μ, the mean number of keys carried among *all* students at your school.

(c) How many of the sample values fall within this interval? What percentage of the students sampled is this? Should this percentage be close to 96%? Explain.

(d) Is this sample technically a simple *random* sample of students at your school? Can you think of any reasons why this sample would not constitute a *representative* sample of students at your school with regard to the issue of number of keys carried? Explain.

Activity 25-7: Exam Score Improvements

The following data are scores on the first and second exams for a sample of students in an introductory statistics course. (The * denotes a missing value; that student did not take the second exam.)

Student	Exam 1	Exam 2	Improvement	Student	Exam 1	Exam 2	Improvement
1	98	80		13	91	92	
2	76	71		14	83	80	
3	90	82		15	83	84	
4	95	68		16	93	96	
5	97	96		17	96	90	
6	89	93		18	98	97	
7	77	50		19	84	95	
8	94	64		20	76	67	
9	88	*	*	21	97	77	
10	95	84		22	72	56	
11	87	76		23	80	78	
12	84	69		24	74	64	

Disregard, for now, the student with the missing value; consider this a sample of 23 students.

(a) Construct (by hand) a dotplot of the *improvements* in scores from exam 1 to exam 2; those who scored lower on Exam 2 than on Exam 1 have negative improvements.

(b) What was the largest improvement from Exam 1 to Exam 2?

(c) What was the biggest decline from Exam 1 to Exam 2?

(d) What proportion of these students scored higher on Exam 1 than on Exam 2?

(e) Use the calculator to compute the sample mean and sample standard deviation of the *improvements*.

(f) Use the calculator to perform the test of whether the mean improvement *differs* significantly from zero. State the null and alternative hypotheses and record the test statistic and p-value.

(g) Based on the significance test conducted in (f), would you reject the null hypothesis at the $\alpha = .10$ significance level? Are the sample data statistically significant at the .05 level? How about at the .01 level?

(h) Use the calculator to find a 95% confidence interval for the mean of the population of *improvements*.

(i) Write a few sentences commenting on the question of whether there seems to have been a significant difference in scores between the two exams.

(j) Now treat the missing value as a score of 0 on the second exam. Use the calculator to recompute the p-value of the test from (f) and the confidence interval in (h).

(k) Comment on whether disregarding the missing value or treating it as a 0 makes much difference in this analysis.

Activity 25-8: Ages of Coins *(cont.)*

Reconsider the data collected in Topic 6 concerning the ages of coins. Select any *one* of the coins (penny, nickel, dime, or quarter) and use the calculator to perform a significance test of whether the mean age of that coin differs significantly from five years. Also have the calculator produce a confidence interval for the mean age of that coin. (Select your own confidence level.) Write a brief paragraph reporting your findings.

Activity 25-9: Marriage Ages *(cont.)*

Recall that the sample of 24 marriage ages that you analyzed in Activity 25-3 is actually a subsample from a larger sample of 100 marriages. Use the calculator to reproduce your analysis of Activity 25-3 using all 100 couples in the larger sample. Write a paragraph reporting on your findings and how they differ from those found by analyzing only 24 couples' ages.

Activity 25-10: Word Lengths *(cont.)*

Reconsider the data that you collected in Topic 1 concerning the numbers of letters in the words of a sentence that you wrote.

(a) Treating these words as a sample from the population of all words that you have written, use the calculator to produce an 80% confidence interval for the mean of this population.
(b) Write a sentence explaining what this interval says.
(c) How many and what proportion of your words in (c) fall within the interval from (a)? Should this proportion be close to 80%? Explain.
(d) How would the interval in (a) have differed if your sample size had been larger (and everything else remained the same)?
(e) How would the interval in (a) have differed if your sample standard deviation had been larger (and everything else remained the same)?
(f) How would the interval in (a) have differed if your sample mean had been smaller (and everything else remained the same)?
(g) How would the interval in (a) have differed if every word in your sample had contained two more letters than it actually did? (Be specific in your answer.)

Activity 25-11: ATM Withdrawals *(cont.)*

Reconsider the ATM withdrawals presented in Activity 4-10. Recall that these data are the amounts of all 111 ATM withdrawals made by a particular individual in 1994.

(a) Use the calculator to look again at visual displays of the distribution of these withdrawal amounts. Also have the calculator compute numerical summaries for the distribution. Write a paragraph commenting on key features of the distribution of withdrawal amounts.

(b) Use the calculator to produce a 96% confidence interval for the population mean μ.

(c) Is there a population in this case from which these 111 withdrawals form a sample? Does the construction of a confidence interval in (b) make any sense? Explain.

Activity 25-12: Students' Family Sizes *(cont.)*

Consider the data collected in Topic 7 concerning students' family sizes.

(a) Find (by hand or with the calculator) a 90% confidence interval for the mean number of siblings among all students at your school. Write a sentence explaining what the interval says.

(b) Determine how many and what proportion of the sample values fall within this interval. Should close to 90% of the students have family sizes which fall within the interval? Explain.

(c) Is this sample a simple random one from the population of all students at your school? If not, is there reason to suspect that the sample is not representative of the population in terms of family size? Explain.

Activity 25-13: Students' Distances from Home *(cont.)*

Consider the data on students' distances from home collected in Topic 4.

(a) Enter these data into the calculator and look at visual displays of the distribution of distances from home. Write a few sentences commenting on key features of the distribution. (As always, remember to relate your comments to the context.)

(b) Have the calculator compute the sample size, sample mean, and sample standard deviation of these distances from home, recording the results.

(c) Let μ denote the mean distance from home among *all* students at your school. Use the calculator to find a 90% confidence interval for μ.

(d) How many of the individual students' distances from home fall within this interval of values? What proportion of the sample is this?

(e) Should you have expected that about 90% of the students sampled would have their distances fall within the 90% confidence interval for μ? Explain.

(f) Have the calculator compute 95% and 99% confidence intervals for the mean distance from home among all students at your school. Record the intervals below, and comment on how they differ from the 90% confidence interval.

(g) List the two formal assumptions which underlie the validity of these confidence intervals, and comment on whether the assumptions seem to be satisfied for these data.

Activity 25-14: Hypothetical Alumni Earnings

Suppose that the Alumni Office of a large university sends out questionnaires to all 5000 members of the graduating class of 1994, asking each to indicate their earnings for their first year out of college. Suppose that the office receives 417 responses and that the mean value reported is $27,952 and that the standard deviation of the values reported is $5477. What can the office assert (with, say, 95% confidence) about the value of μ, the mean first-year earnings of *all* members of the class of 1994? Explain.

Activity 25-15: Properties of Intervals

(a) For a fixed sample size and confidence level, what happens to the half-width of a confidence interval for a population mean μ as the sample standard deviation increases?

(b) For a fixed sample size and sample standard deviation, what happens to the half-width of a confidence interval for a population mean μ as the confidence level increases?

(c) For a fixed confidence level and sample standard deviation, what happens to the half-width of a confidence interval for a population mean μ as the sample size increases?

(d) For a fixed sample size, confidence level, and sample standard deviation, what happens to the half-width of a confidence interval for a population mean μ as the sample mean increases?

Activity 25-16: Confidence Interval of Personal Interest (*cont.*)

Think of a real situation in which you would be interested in producing a confidence interval to estimate a population *mean*. Describe precisely the population and parameter involved. Also describe how you might select a sample from the population. (Be sure to think of a *measurement* variable so that a mean is a sensible parameter to deal with.)

WRAP-UP

This topic has asked you to apply techniques of inference about a population mean to a variety of problems. In addition to practicing the mechanics of the calculations involved with the inference procedures, you have also been asked to consider proper and improper interpretations and uses of the procedures.

The new idea in this topic is the *paired comparisons* design, which allows an experimenter to control for possible effects of extraneous variables. The analysis of data from such a design involves applying inference procedures for a population mean to the sample of differences.

The next topic will continue to explore inference concerning means but will move to the question of comparing means from two independent groups. Once again, you will find that the reasoning, structure, and interpretation of the inference procedures is unchanged.

Topic 26:

COMPARING TWO MEANS

OVERVIEW

You have been studying the application of inference techniques to various situations involving genuine data. In the previous two topics you have investigated problems which call for inferences about a population mean. With this topic you will examine the case of comparing two sample means where the samples have been collected *independently* (as opposed to the paired comparisons design that you studied in the last topic). The inference procedures will again be based on the *t*-distribution; the reasoning behind and interpretation of the procedures remain the same as always. Also as always, you will see the importance of an initial examination of the data, visual and numerical, prior to applying formal inference procedures.

OBJECTIVES

- To develop a sense of the factors that are important to consider when comparing two sample means.
- To learn a significance testing procedure for determining whether two sample means differ "significantly" and to understand the reasoning behind and limitations of the procedure.
- To learn a confidence interval procedure for estimating the magnitude of the difference between two population means.
- To explore the effects of the various factors that affect inference procedures concerning a difference in population means.
- To acquire proficiency using graphical, descriptive, and inferential methods to analyze data concerned with questions of whether a difference exists between two groups or two treatments.

PRELIMINARIES

1. Do you suspect that women tend to pay more for haircuts than men do?

2. Do you think that *every* woman pays more for her haircut than *every* man does?

3. By how much (if any) do you think women tend to outspend men for a haircut on the average?

4. Mark on the scale below an interval that you believe with 95% confidence to contain the difference in mean haircut prices paid by women and men at your school (women's mean price minus men's mean price, in dollars).

```
+---------+---------+---------+---------+---------+-
-6.0      -3.0       0.0       3.0       6.0       9.0
```

5. Record below the prices of the most recent haircut for the students in your class; also record whether the student is a man or a woman.

Student	Gender	$ haircut	Student	Gender	$ haircut	Student	Gender	$ haircut
1			9			17		
2			10			18		
3			11			19		
4			12			20		
5			13			21		
6			14			22		
7			15			23		
8			16			24		

6. Do you suspect that second grade instructors spend more individualized instructional time with boys or with girls when teaching *mathematics*, or do you not expect to see a gender difference there?

7. Do you suspect that second grade instructors spend more individualized instructional time with boys or with girls when teaching *reading*, or do you not expect to see a gender difference there?

IN-CLASS ACTIVITIES

Activity 26-1: Hypothetical Commuting Times

Suppose that a commuter Alex wants to determine which of two possible driving routes gets him to work more quickly. Suppose that he drives route 1 for 10 days and route 2 for 10 days, recording the commuting times (in minutes) and displaying them as follows:

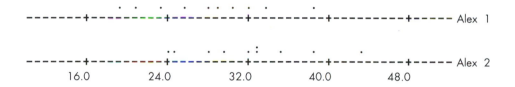

(a) For this sample of days, does one route *always* get Alex to work more quickly than the other?

(b) For this sample of days, does one route *tend* to get Alex to work more quickly than the other? If so, identify the route which seems to be quicker.

(c) Hypothetical commuter times have been stored in a grouped file named HYPOCOMM.83g. Download this grouped file into your calculator and determine the sample means and sample standard deviations of these commuting times for each route; record them in the table below:

	Sample size	Sample mean	Sample std. dev.
Alex 1			
Alex 2			

Because of sampling variability, one cannot conclude that because these sample means differ, the means of the respective populations must differ as well. As always, one can use a test of significance to establish whether a sample result (in this case, the observed difference in sample mean commuting times) is "significant" in the sense of being unlikely to have occurred by chance alone. Also as always, one can use a confidence interval to estimate the magnitude of the difference in population means.

Inference procedures for comparing the population means of two different groups are similar to those for comparing population proportions in that they take into account sample information from both groups; they are similar to inference procedures for a single population mean in that the sample sizes, sample means, and sample standard deviations are the relevant summary statistics. These statistics are denoted by the following notation:

	First group	Second group
Sample size	n_1	n_2
Sample mean	$\bar{x}_1$	$\bar{x}_2$
Sample standard deviation	s_1	s_2

The forms for confidence intervals and significance tests concerning the difference between two population means, which will be denoted by $\mu_1 - \mu_2$, are presented below. (The use of the subscripts will indicate the population from which the measurements come.) The important point to remember is that the reasoning behind and interpretation of these procedures is always the same.

Confidence interval for $\mu_1 - \mu_2$:

$$(\bar{x}_1 - \bar{x}_2) \pm t_k^* \sqrt{\frac{s_1^2}{n_1} + \frac{s_2^2}{n_2}}$$

where t_k^* is the appropriate critical value from the t-distribution with degrees of freedom k equal to the smaller of $n_1 - 1$ and $n_2 - 1$.

Significance test of equality of μ_1 and μ_2:

$$H_0 : \mu_1 = \mu_2$$
$$H_a : \mu_1 < \mu_2, \quad \text{or} \quad H_a : \mu_1 > \mu_2, \quad \text{or} \quad H_a : \mu_1 \neq \mu_2$$

Test statistic: $t = \dfrac{\bar{x}_1 - \bar{x}_2}{\sqrt{\dfrac{s_1^2}{n_1} + \dfrac{s_2^2}{n_2}}}$

p-value $= \Pr(T_k < t), \text{ or } \Pr(T_k > t), \text{ or } 2\,\Pr(T_k > |t|)$

where T_k represents a t-distribution with degrees of freedom equal to the smaller of $n_1 - 1$ and $n_2 - 1$.

Assumptions:

1. The two samples are independently selected simple random samples from the populations of interest.
2. *Either* both sample sizes are large ($n_1 \geq 30$ and $n_2 \geq 30$ as a rule-of-thumb) *or* both populations are normally distributed.

(d) Use the 2-SampTTest feature (located in the STAT TESTS menu of the calculator) to conduct a significance test of whether Alex's sample commuting times provides evidence that the mean commuting times with these two routes differ. Record the p-value of the test in the table below, along with the sample statistics that you have already calculated.

	Sample size	Sample mean	Sample std. dev.	Two-sided p-value
Alex 1				
Alex 2				

(e) Are Alex's sample results statistically significant at any of the commonly used significance levels? Can Alex reasonably conclude that one route is faster than the other for getting to work? Explain.

The relatively large p-value (0.15) of this test reveals that the sample commuting times do *not* constitute strong evidence that one route tends to be quicker than the other.

Consider now three other commuters who conduct similar experiments to compare travel times for two different driving routes. For the sake of comparison, Alex's results are also reproduced below:

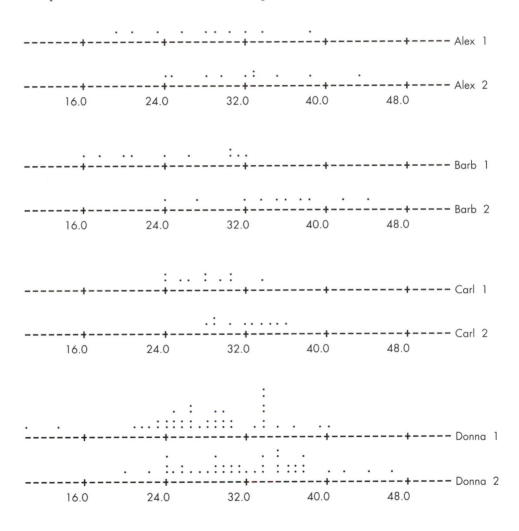

(f) Based on your visual analysis of these pairs of dotplots, what strikes you as the most important difference between Alex's and Barb's results?

(g) Based on your visual analysis of these pairs of dotplots, what strikes you as the most important difference between Alex's and Carl's results?

(h) Based on your visual analysis of these pairs of dotplots, what strikes you as the most important difference between Alex's and Donna's results?

(i) For each commuter (Barb, Carl, and Donna), use the calculator to conduct a significance test of whether the difference in his or her sample mean commuting times differ significantly. Record the p-values of these tests below, along with the appropriate sample statistics.

	Sample size	Sample mean	Sample std. dev.	Two-sided p-value
Barb 1				
Barb 2				
Carl 1				
Carl 2				
Donna 1				
Donna 2				

(j) For each of Barb, Carl, and Donna, explain why his or her sample results differed significantly while Alex's did not.

Barb:

Carl:

Donna:

Activity 26-2: Students' Haircut Prices

Use the data collected in the "Preliminaries" to investigate the proposition that women tend to pay more for their haircuts than men do.

(a) Construct (by hand, on the same scale) dotplots of the distribution of haircut prices paid by men and by women.

(b) Did *every* woman in the sample pay more for her haircut than did every man in the sample? Does it seem as though *most* women paid more than *most* men for their haircuts? Does it seem as though women paid more *on the average* than men for their haircuts? Explain.

(c) Enter these data into the calculator and have the calculator compute summary statistics for both samples; record the results in the following table.

	Sample size	Sample mean	Sample std. dev.
Men			
Women			

(d) Let μ_w denote the mean price paid by all women at your school for their most recent haircut, and let μ_m denote the respective mean for men. Conduct (by hand) the appropriate test of significance to assess whether the average price paid by *all* women for their most recent haircut exceeds that of men. Please report the null and alternative hypotheses, and show the calculations of the test statistic and p-value. Then write a one-sentence conclusion about the question of interest.

(e) Find (by hand) a 96% confidence interval for the difference in mean price of haircuts between men and women. Then comment on what this confidence interval reveals about the question of interest.

Activity 26-3: Trading for Run Production

In the first 94 games of the Atlanta Braves' 1993 season, the team scored an average of 3.99 runs per game. At that point the Braves traded for Fred McGriff, a player renowned for his offensive power. The team hoped that McGriff would increase their run production in the remainder of the season. Over the course of those remaining 68 games, the Braves scored an average of 5.78 runs per game.

(a) What further information would you need to be able to conduct a test of significance to determine whether this difference in sample means is unlikely to have occurred by chance alone?

(b) The raw data (runs scored in each of the 162 games) appear below.

Without				With		
1	3	7	3	8	4	1
0	3	3	3	14	1	3
5	4	3	8	7	3	13
6	3	2	6	6	5	4
2	1	1	0	11	5	10
1	13	5	6	13	3	7
3	13	4	3	12	6	3
1	8	4	4	10	4	2
3	12	1	2	3	5	2
0	2	2	11	0	6	11
1	5	4	4	4	9	18
0	5	4	9	4	7	1
2	10	1	4	3	6	6
12	5	6	1	3	8	0
5	4	7	1	9	8	9
7	5	9	5	4	2	7
3	0	4	2	2	5	2
3	1	2	6	3	7	6
11	4	5	4	3	3	8
3	1	1	3	3	3	7
3	2	4	3	4	1	10
2	5	5	2	8	1	5
1	0	8	0	14	8	
5	4					

The data in the table above has been stored in a grouped file named MCGRIFF.83g. Download this grouped file into your calculator. Use the calculator to analyze the raw data (runs scored in each of the 162 games) by calculating five-number summaries of the distributions of runs scored with McGriff and without McGriff. Record these in the table below. Also have the calculator generate boxplots (on the same scale) of the two distributions. Comment on what the boxplots reveal about the two distributions with regard to the question of whether McGriff seemed to increase the Braves' run production.

	Minimum	Lower quartile	Median	Upper quartile	Maximum
Without McGriff					
With McGriff					

(c) Have the calculator compute the sample sizes, sample means, and sample standard deviations of the Braves' runs scored with McGriff and without McGriff.

	Sample size	Sample mean	Sample std. dev.
Without McGriff			
With McGriff			

(d) Use the calculator to conduct the appropriate test of whether the increase in sample mean runs scored is statistically significant at the $\alpha = .05$ level. Report the test statistic and p-value.

(e) Is the increase in sample mean runs scored likely to have occurred by chance alone? Explain.

(f) Does this analysis allow you to conclude that McGriff was the sole cause of the Braves' improved run production after the trade? Explain, being sure to mention whether this is a controlled experiment or an observational study.

HOMEWORK ACTIVITIES

Activity 26-4: Hypothetical Commuting Times (*cont.*)

Reconsider the commuting time experiments presented in Activity 26-1. For each of the following people (Earl, Fred, Grace, Harry, and Ida), determine whether or not his or her sample results are statistically significant. Do *not* perform any calculations; simply compare the sample results to those of Alex, Barb, Carl, or Donna. In each case indicate whose results you use for the comparison and explain your answer.

	Sample size	Sample mean	Sample std. dev.
Earl 1	10	29.0	6.0
Earl 2	10	31.0	6.0
Fred 1	75	28.0	6.0
Fred 2	75	32.0	6.0
Grace 1	10	20.0	6.0
Grace 2	10	40.0	6.0
Harry 1	10	28.0	9.0
Harry 2	10	32.0	9.0
Ida 1	10	28.0	1.5
Ida 2	10	32.0	1.5

Activity 26-5: Lengths and Scores of Baseball Games

An undergraduate researcher recorded the lengths (in minutes) and scores of the Major League baseball games played in June 1992. In questions (a)–(e), ignore the distinction between National and American League games. The data appear below:

Win	Lose	Leag	Time	Win	Lose	Leag	Time	Win	Lose	Leag	Time	Win	Lose	Leag	Time
8	6	NL	202	16	3	AL	203	7	5	AL	170	7	1	AL	170
7	6	NL	187	5	2	AL	155	4	3	AL	220	2	1	AL	144
6	1	NL	133	6	2	AL	152	9	6	AL	177	5	3	AL	146
14	1	NL	198	9	2	AL	211	2	1	NL	137	11	7	AL	212
7	1	NL	186	5	1	AL	165	5	3	NL	148	9	6	NL	198
6	2	AL	165	2	1	AL	183	5	2	NL	172	4	1	NL	170
5	3	AL	154	6	1	AL	197	9	8	NL	220	3	2	NL	146
5	3	AL	197	4	2	AL	158	11	0	NL	177	8	1	NL	170
7	1	AL	179	2	1	AL	167	6	5	NL	249	5	0	NL	135
8	2	AL	209	4	2	AL	160	4	2	AL	173	11	0	AL	135
10	7	AL	120	6	1	AL	157	6	2	AL	195	8	4	AL	201
3	2	NL	202	4	1	AL	171	3	2	AL	141	7	2	AL	140
1	0	NL	143	5	3	NL	193	10	2	AL	161	4	3	AL	187
5	3	NL	198	5	4	NL	152	4	3	AL	178	6	3	AL	170
4	3	NL	190	6	5	NL	199	2	1	AL	131	5	1	AL	164
2	1	NL	150	3	2	NL	161	3	0	AL	130	3	2	AL	194
6	0	NL	142	3	2	NL	150	5	0	NL	142	8	5	NL	202
4	3	AL	141	6	5	NL	196	4	3	NL	155	9	2	NL	142
2	1	AL	146	5	2	AL	167	6	4	NL	177	8	0	NL	149
8	2	AL	155	10	3	AL	204	8	2	NL	145	5	2	AL	152
5	4	AL	183	3	2	AL	248	5	2	NL	187	4	3	AL	180
7	5	AL	254	3	1	AL	179	3	1	AL	138	13	4	AL	177
4	2	AL	135	4	2	AL	217	5	4	AL	209	1	0	AL	139
5	4	AL	168	7	6	AL	173	5	4	AL	246	5	1	AL	172
5	3	NL	186	5	3	AL	178	4	1	AL	150	6	2	NL	146
8	7	NL	160	6	2	NL	142	14	10	AL	241	4	3	NL	194
4	1	NL	162	5	1	NL	149	9	4	NL	174	6	2	NL	170
5	1	NL	164	2	1	NL	195	8	3	NL	211	3	0	NL	142
6	5	NL	189	2	1	NL	175	7	5	NL	209	6	5	NL	205
8	3	AL	159	4	2	NL	144	4	3	NL	181	7	4	NL	136
11	3	AL	184	8	2	NL	204	4	0	NL	150	6	1	AL	152
3	1	AL	131	7	5	AL	201	8	3	AL	258	8	4	AL	160
4	3	AL	228	4	0	AL	178	12	8	AL	196	6	5	AL	170
10	4	AL	199	4	0	AL	163	10	7	AL	200	4	3	AL	197
7	6	AL	210	6	0	AL	180	4	1	AL	170	2	1	AL	181
4	3	AL	271	6	5	AL	308	11	4	AL	173	4	2	AL	177
15	12	AL	240	5	1	AL	164	1	0	AL	141	10	1	AL	187
10	3	AL	160	5	0	AL	169	5	3	AL	191	12	3	AL	121
6	2	AL	213	3	0	AL	120	2	1	NL	233	2	1	NL	227
7	2	NL	171	4	1	AL	125	3	2	NL	175	5	4	NL	169
12	6	NL	180	5	0	AL	141	2	1	NL	144	5	1	NL	149
7	4	NL	172	3	2	NL	168	4	3	NL	200	12	4	NL	200
5	4	NL	209	4	3	NL	186	5	2	NL	165	5	3	NL	185
7	5	NL	182	5	2	NL	166	3	2	NL	182	10	8	AL	188
3	2	NL	161	8	5	NL	183	4	1	AL	157	8	7	AL	233
6	2	NL	175	6	4	NL	205	5	3	AL	166	2	1	AL	162
5	4	NL	199	3	2	NL	163	10	0	AL	162	12	2	AL	191
10	4	NL	175	15	1	AL	191	3	1	AL	171	6	4	AL	174

(continues)

Win	Lose	Leag	Time	Win	Lose	Leag	Time	Win	Lose	Leag	Time	Win	Lose	Leag	Time
6	2	NL	137	5	3	AL	182	4	1	AL	167	8	7	AL	250
6	4	AL	157	7	1	AL	154	9	5	AL	199	2	0	AL	146
5	4	AL	166	4	1	AL	160	6	1	AL	158	3	2	NL	219
1	0	AL	152	5	4	AL	167	1	0	NL	151	7	3	NL	168
7	1	AL	159	4	2	AL	192	3	1	NL	183	8	2	NL	188
10	3	AL	197	8	7	AL	201	2	1	N;	130	6	5	NL	165
4	3	AL	150	11	1	NL	161	4	1	NL	134	9	0	NL	147
7	5	NL	176	4	2	NL	150	4	3	NL	146	5	3	NL	137
12	6	NL	204	4	3	NL	137	6	1	NL	147	9	3	AL	192
5	4	NL	200	4	1	NL	188	2	0	AL	152	7	6	AL	194
15	1	NL	196	3	2	NL	227	8	2	AL	177	9	2	AL	179
5	1	NL	145	7	4	AL	183	4	2	AL	173	8	4	AL	215
6	4	AL	171	6	1	AL	160	6	5	AL	206	10	2	AL	171
6	1	AL	179	6	2	AL	184	5	0	AL	150	9	2	AL	153
4	3	AL	187	4	3	AL	190	3	2	AL	187	6	3	AL	183
6	2	AL	175	8	7	AL	178	4	2	AL	170	6	5	NL	238
4	3	AL	154	5	1	AL	146	5	2	NL	154	5	4	NL	164
5	1	AL	169	14	4	AL	169	6	2	NL	169	3	1	NL	134
3	1	AL	144	5	1	NL	158	2	0	NL	156	5	2	NL	154
5	4	NL	185	4	2	NL	150	1	0	NL	163	4	3	NL	169
1	0	NL	144	4	1	NL	159	2	0	NL	153	5	3	AL	156
3	0	NL	164	5	2	NL	174	5	4	NL	203	7	3	AL	156
9	4	NL	151	5	1	NL	156	16	7	AL	228	5	4	AL	220
3	0	NL	134	5	4	NL	187	7	2	AL	175	5	1	AL	141
3	2	NL	173	15	7	NL	186	4	2	AL	211	11	4	AL	213
10	3	AL	185	6	5	AL	211	7	1	AL	140	9	6	AL	192
5	4	AL	154	3	2	AL	199	2	0	AL	146	8	3	AL	204
6	5	AL	195	5	2	AL	153	5	4	AL	167	5	4	AL	201
4	1	AL	150	4	1	AL	176	8	2	NL	209	12	3	AL	189
6	1	AL	149	1	0	AL	137	5	3	NL	206	16	13	AL	225
4	0	AL	174	7	0	AL	172	5	2	NL	146	2	0	AL	143
7	1	AL	177	7	1	NL	121	5	2	NL	138	4	2	AL	197
4	2	NL	173	4	1	NL	182	4	2	NL	178	6	0	AL	135
4	1	NL	153	4	1	NL	148	8	4	NL	162	8	5	AL	185
3	2	NL	153	7	5	NL	158	6	4	NL	167	4	3	NL	150
7	0	NL	147	7	1	NL	175	10	6	NL	181	7	2	NL	174
6	0	NL	157	2	0	NL	137	7	0	NL	176	3	1	NL	158
5	2	NL	261	10	0	AL	151	5	0	NL	158	2	0	NL	133
6	4	NL	201	7	5	AL	177	4	1	NL	171	5	1	NL	185
9	6	AL	164	4	3	AL	188	7	1	AL	162	2	1	NL	176
14	3	AL	174	4	1	AL	142	12	7	AL	172				

(a) The lengths and scores of the Major League baseball games played in June of 1992 have been stored in a grouped file named BASEBALL.83g. Download this grouped file into your calculator. Use the calculator to produce visual displays and numerical summaries of the game lengths. Write a paragraph commenting on key features of this distribution.

(b) Use the calculator to produce a 95% confidence interval for μ, the mean length of all Major League baseball games played in 1992. Write a sentence explaining what the interval says.

(c) Would you expect that about 95% of the games played that season had lengths that fall within this interval? Explain.

(d) Repeat (a), (b), and (c) for the total runs scored in a game.

(e) Repeat (a), (b), and (c) for the margin of victory in a game.

(f) Use the calculator to test whether the sample games provide strong evidence that the mean lengths of games differ between the National and American Leagues. Also have the calculator produce a 95% confidence interval to estimate the difference in mean game lengths. Report below the null and alternative hypotheses, the test statistic, the p-value, and the confidence interval. Finally, write a short paragraph summarizing your findings on the comparison of game lengths between the two leagues.

(g) Repeat (f) with regard to comparing total runs scored between the two leagues.

(h) Repeat (f) with regard to comparing margin of victory between the two leagues.

Activity 26-6: Lifetimes of Notables (*cont.*)

Refer to Activity 6-8, where you analyzed distributions of lifetimes of "noted personalities." Recall that the categories of notables are the following: scientists, writers, politicians, military leaders, artists, philosophers, social reformers, historians, and business leaders. Summary statistics appear below.

Notable	Sample size	Sample mean	Sample std. dev.
Scientists	133	73.04	13.13
Writers	302	65.65	15.20
Politicians	232	68.49	14.09
Military leaders	101	66.27	15.45
Artists	205	69.02	15.38
Philosophers	91	69.20	14.13
Social reformers	51	75.39	12.33
Historians	94	72.59	11.93
Business leaders	101	76.31	9.34

(a) Pick out any *two* of these groups of notables that are of interest to you. List your choices and designate one as group 1 and the other as group 2.

(b) Perform the significance test to assess whether the data provide evidence that the means of *populations* truly differ. Be sure to state the null and alternative hypotheses (identifying whatever symbols you introduce) and to report the test statistic and p-value. Also state exactly what the p-value says in this context, and write a one-sentence conclusion.

(c) Find a 95% confidence interval for the difference in population means.

(d) Write a sentence or two interpreting the interval that you've found in part (c). In particular, comment on whether the interval contains only negative values, only positive values, or some of both. Also indicate what this says about the mean lifetimes of the populations represented by your two groups.

(e) Indicate specifically how results in part (b) and (c) would have differed if you had labeled your two groups in the opposite manner (i.e., if the group you labeled as group 1 had been labeled group 2, and vice versa.)

(f) List the assumptions that underlie the validity of thses procedures and comment on whether they seem to be satisified. In particular, do you need to assume anything about 18 the distributions of individual lifetimes?

(g) Use the calculator to determine a 95% confidence interval for the population mean lifetime of the first group that you chose. Then do the same for your second group of notables.

(h) Determine and record the half-widths of the three confidence intervals that you have found (i.e., for the difference in population means, for the population mean of group 1, and for the population mean of group 2).

(i) Compare these half-widths. Specificallly, is the half-width of the interval estimating the difference in population means larger than either of the individual group half-widths? Is the half-width of the interval estimating the difference in population means as large as the sum of the two individual group half-widths?

Activity 26-7: Word Lengths *(cont.)*

Consider the word lengths that you analyzed in Activity 6-14.

(a) Use the calculator to create visual displays comparing your distribution of word lengths with mine. Comment on similarities and differences that you observe in these distributions.

(b) Have the calculator compute summary statistics for these two distributions of word lengths. Record the sample sizes, sample means, and sample standard deviations.

(c) Use these sample statistics to conduct (by hand) a test of significance to assess whether these sample data provide evidence that your mean number of letters per word differs from mine. Report the null and alternative hypotheses (identifying whatever symbols you introduce), and show the calculations of the test statistic and p-value. Finally, write a one-sentence conclusion about the question of interest, i.e., whether the data provide evidence that our means of letters per word differ.

Activity 26-8: Ages of Coins *(cont.)*

Recall the data from Topic 6 concerning ages of coins. For each of the six pairs of coins (penny-nickel, penny-dime, etc.), have the calculator conduct a test of whether their mean ages differ significantly. Record the test statistic and p-value for each test as well as the sample means and standard deviations for each coin. Write a few sentences summarizing your findings.

Activity 26-9: Classroom Attention

Researchers in a 1979 study recorded the lengths of individual instructional time (in seconds) that second grade instructors spent with their students. They compared these times between girls and boys in the subjects of reading and mathematics. Numerical summaries of their results appear below:

Reading	Sample size	Sample mean	Sample std. dev.
Boys	372	35.90	18.46
Girls	354	37.81	18.64

Mathematics	Sample size	Sample mean	Sample std. dev.
Boys	372	38.77	18.93
Girls	354	29.55	16.59

Conduct (either by hand or with the calculator) appropriate tests of significance to indicate whether the sample mean instructional times differ significantly between boys and girls in either subject. Also produce confidence intervals for the difference in population means between boys and girls. Write a paragraph or two summarizing your conclusions.

Activity 26-10: UFO Sighters' Personalities *(cont.)*

Reconsider the study described in Activity 21-7, (part of the data from which you analyzed in Activity 24-11.) A control group of 53 community members (who had not reported UFO experiences) had a mean IQ of 100.6 with a standard deviation of 12.3. Use this information together with that presented in Activity 24-11 to test whether the mean IQ's of community members and UFO sighters differ significantly. Show the details of the test and write a one-sentence conclusion. Also find a confidence interval for the difference in population means and explain what the interval reveals.

Activity 26-11: Students' Measurements *(cont.)*

Consider once again the data collected in Topic 3 on foot lengths, heights, and armspans. Choose one of these three variables.

(a) Use the calculator to test whether men's and women's means differ significantly at the .05 level. Report the details of the test procedure and write a brief conclusion.

(b) Use the calculator to produce a 97.5% confidence interval to estimate the difference in population means between men and women. Write a sentence explaining what the interval reveals.

(c) How would the interval in (b) have differed if the sample sizes had been larger (and everything else had been the same)?

(d) How would the interval in (b) have differed if the sample means had been closer together (and everything else had been the same)?

(e) How would the interval in (b) have differed if the sample standard deviations had been larger (and everything else had been the same)?

(f) List the technical assumptions necessary for these procedures to be valid in this case, and then comment on whether the assumptions seem to be satisfied.

Activity 26-12: Mutual Fund Returns *(cont.)*

Recall the data from Activity 6-15 concerning the percentage returns of mutual funds which charge load fees and of those which do not have loads. Use the calculator to analyze these data in an appropriate manner to address the question of whether funds with loads tend to outperform those without loads and, if so, by about how much. Write a paragraph or two detailing your findings.

Activity 26-13: Tennis Simulations *(cont.)*

Refer again to the data of Activities 2-4, 4-9, and 6-6 concerning three different scoring systems for tennis.

(a) Use the calculator to analyze whether the mean game length with conventional scoring differs significantly from the mean game length with no-ad scoring and, if so, by about how much. Write a paragraph describing your findings.
(b) Repeat (a), comparing no-ad scoring with handicap scoring.

Activity 26-14: Marriage Ages *(cont.)*

Reconsider the data on marriage ages of 100 couples from Activity 25-9. Since these data are *paired*, you were asked in that activity to perform the correct analysis on the *differences* in marriage ages. Suppose now (incorrectly) that the data had been independent samples—one sample of 100 husbands' ages and another sample of 100 wives' ages. The sample statistics are:

	Sample size	Sample mean	Sample std. dev.
Husband	100	33.08	12.31
Wife	100	31.16	11.00

(a) Use this information to conduct (by hand) a two-sample test (not a paired test) of whether the sample mean age for husbands exceeds that of wives by a statistically significant margin. Report the details of the test procedure and write a one-sentence conclusion.

(b) Find (by hand) a 95% confidence interval for the difference in mean ages between husbands and wives in the population. (Continue to assume that these are two independent samples.)

(c) Comment on how these results differ from those of Activity 24-9 where you performed the appropriate paired analysis of these data. Also comment on how these differences highlight the usefulness of the paired comparisons design.

Activity 26-15: Hypothetical ATM Withdrawals *(cont.)*

Reconsider the hypothetical ATM withdrawals first presented in Activity 5-18. In Activity 25-2 you calculated sample sizes, sample means, and sample standard deviations of the withdrawal amounts for each machine.

(a) Choose any pair of machines and calculate (by hand, using the summary statistics from Activity 25-2) a 90% confidence interval for estimating the difference in population mean withdrawal amounts between the two machines.

(b) Does this interval include the value zero?

(c) Would this interval be any different if you had chosen a different pair of machines? Explain.

(d) Are these three machines identical in their distributions of withdrawal amounts? Explain.

Activity 26-16: Hypothetical Bowlers' Scores

Suppose that three bowlers Chris, Fran, and Pat bowl 36 games each with the following results:

Chris	132	135	136	140	140	142	142	143	144	147	147	148
	149	149	150	152	153	154	155	155	156	158	158	159
	160	161	161	162	162	163	165	165	167	169	169	170
Fran	72	76	79	88	93	99	104	170	171	172	172	172
	173	173	174	174	175	175	175	176	178	178	179	179
	181	183	183	184	185	186	186	186	187	191	191	194
Pat	170	171	172	172	172	173	173	173	174	174	174	175
	175	175	176	176	178	178	178	179	179	179	181	183
	183	183	184	185	186	186	186	187	188	191	191	194

(a) Use the calculator to produce dotplots of the three bowlers' scores on the same scale. Write a few sentences comparing and contrasting the three distributions of scores.

Consider these scores as random samples from the hypothetically infinite population of scores that these bowlers would achieve in the long run. Let μ_C, μ_F, and μ_P represent the population means of the three respective bowlers.

(b) Have the calculator perform the appropriate significance test of whether μ_C differs from μ_F. Report the test statistic and p-value of the test. Is the difference in sample means statistically significant at any of the common significance levels?

(c) Now have the calculator perform the appropriate significance test of whether μ_F differs from μ_P. Report the test statistic and p-value of the test. Is the difference in sample means statistically significant at any of the common significance levels?

(d) Comment on how this example illustrates that one should look at an entire distribution of values rather than just concentrate on the mean. That is, what would you have missed about the bowlers if you had performed the significance tests without looking at the data first?

Activity 26-17: Hypothetical Bowlers' Scores *(cont.)*

Suppose two bowlers Jack and Jill each bowl 1000 games. Suppose further that Jack's scores have a mean of 166.8 and a standard deviation of 14.9, while Jill's scores have a mean of 169.3 and a standard deviation of 14.6.

(a) Use this sample information to determine whether Jack's and Jill's population means differ significantly at the $\alpha = .01$ level. Report the null and alternative hypotheses, and show the calculations of the test statistic and p-value.

(b) Construct a 95% confidence interval for the difference in Jack's and Jill's population means. Show the details of your calculation, and comment on whether or not the interval contains the value zero.

(c) Which of the following statements is more accurate:

- The sample data provide extremely strong evidence that Jill has a higher mean bowling score than does Jack.
- The sample data provide strong evidence that Jill has a substantially higher mean bowling score than does Jack.

Also explain the subtle difference between these two statements.

Activity 26-18: Students' Haircut Prices *(cont.)*

Reconsider the data collected above on students' haircut prices.

(a) Find (by hand, using the sample statistics computed in Activity 26-2) a 98% confidence interval for the mean price paid by *men* at your school for their most recent haircut.

(b) Find (again by hand, using the sample statistics computed in Activity 26-2) a 98% confidence interval for the mean price paid by *women* at your school for their most recent haircut.

(c) Mark these two intervals on the same scale. Do they overlap?

(d) Write a sentence or two commenting on what these intervals reveal about the mean prices paid by college men and women for their haircuts.

Activity 26-19: Inference of Personal Interest

The following wrap-up asserts that "the underlying principle of all statistical inference techniques is that one uses *sample statistics* to learn something (i.e., to *infer* something) about *population parameters.*" Write a paragraph demonstrating that you understand this statement by describing a situation in which you would be interested in using a sample statistic to infer something about a population parameter. Clearly identify the sample, population, statistic, and parameter in your example. Be as specific as possible, and do not use any example which appears in this text.

WRAP-UP

This topic has extended your knowledge of inference techniques to include the goal of comparing two means. You have discovered the role that the sample sizes and sample standard deviations play in the inference procedures, and you have applied the procedures to a host of genuine examples.

In the last few units, you have learned how to apply inference procedures to a variety of situations involving population means, comparing two population means, population proportions, and comparing two population proportions. As you would expect, there are many more inference techniques to apply to situations other than the ones you have studied. For example, the commonly used technique of *analysis of variance* allows one to compare the means of sample observations from more than two groups simultaneously. *Chi-square tests* enable one to detect relationships between categorical variables as summarized in two-way tables, and *regression analysis* also makes use of inferential techniques.

The essential point that these units have repeatedly emphasized is that while the details of the calculations vary from situation to situation, the reasoning, structure, and interpretation of confidence intervals and of significance tests (the two major types of classical statistical inference) never change. If you have acquired a fundamental understanding of what these inference procedures do (and do not) mean, you should have little difficulty in applying and interpreting them in other circumstances.

Another message that these final units on applications of inference have concentrated on is that the application of formal inference techniques is not a substitute for the exploratory graphical and numerical techniques that you learned in the first two units. It is always a good idea to look at visual displays of a set of data that you are analyzing *before* proceeding with formal inference procedures. Yet another theme has been that questions of sampling and experimental design are also essential for applying and interpreting inferential techniques properly.

Of course, the underlying principle of all statistical inference techniques is that one uses *sample statistics* to learn something (i.e., to *infer* something) about *population parameters*. All of these inference procedures are based on the assumption that the data collected are a simple random sample from the population of interest; it is always important to ask whether this assumption is satisfied before drawing inferences and interpreting their results.

One common (and very legitimate) concern of students studying these various inference techniques is the ability to recognize which situation a certain problem falls in. One guide in addressing this problem is to ask yourself the following series of questions:

1. Are the variables *measurement* or *categorical* ones; i.e., do the data consist of measurements or of "yes-no" type responses? If they are measurements (like sleeping times, or distances from home, or SAT scores), then one typically considers *means*; if they are yes-no responses (like whether or not a spun penny lands heads or whether or not a job applicant passes an employment test), then one typically considers *proportions*.
2. Is there just *one* population of interest, or are *two* populations being *compared*? For example, are you interested in the mean sleeping time of college students or in comparing the mean sleeping times of men and women? Are you interested in the proportion of people who catch a cold during the winter or in comparing the proportions who catch a cold between a group that took vitamin C and a group that took a placebo?
3. If there are two groups and means are of interest, is the design one of *paired comparisons* or of *independent samples*? For example, did one sample of baseball players run a wide angle around first base and another independent sample of baseball players take a narrow angle, or did every player run both angles so that *differences* in times for each player could be analyzed?

For your studying convenience, the following pages review the specific details of all of the inference procedures that you have used:

Inference Procedures for a Single Population Proportion:

Confidence interval for a population proportion θ:

$$\hat{p} \pm z^{\star} \sqrt{\frac{\hat{p}(1 - \hat{p})}{n}},$$

where $z^{\star}$ is the appropriate critical value from the standard normal distribution.

Significance test about a population proportion θ:

$$H_0 : \theta = \theta_0$$
$$H_a : \theta < \theta_0, \quad \text{or} \quad H_a : \theta > \theta_0, \quad \text{or} \quad H_a : \theta \neq \theta_0$$

$$\text{Test statistic: } z = \frac{\hat{p} - \theta_0}{\sqrt{\frac{\theta_0(1 - \theta_0)}{n}}}$$

p-value: $\Pr(Z < z)$, or $\Pr(Z > z)$, or $2 \Pr(Z > |z|)$

where θ_0 represents the hypothesized value of the population proportion and Z represents the standard normal distribution.

Assumptions:

1. The sample is a simple random sample (SRS) from the population of interest.
2. The sample size is large ($n \geq 30$ as a rule-of-thumb).

Calculator Features:

1. 1-PropZInt
2. 1-PropZTest

Inference Procedures for Comparing Two Population Proportions:

Confidence interval for a difference in population proportions $\theta_1 - \theta_2$:

$$(\hat{p}_1 - \hat{p}_2) \pm z^{\star} \sqrt{\frac{\hat{p}_1(1 - \hat{p}_1)}{n_1} + \frac{\hat{p}_2(1 - \hat{p}_2)}{n_2}},$$

where $z^{\star}$ is the appropriate critical value from the standard normal distribution.

Significance test of equality of two population proportions θ_1 and θ_2:

$$H_0 : \theta_1 = \theta_2$$
$$H_a : \theta_1 < \theta_2, \quad \text{or} \quad H_a : \theta_1 > \theta_2, \quad \text{or} \quad H_a : \theta_1 \neq \theta_2$$

Test statistic: $z = \dfrac{\hat{p}_1 - \hat{p}_2}{\sqrt{\hat{p}_c(1 - \hat{p}_c)(\dfrac{1}{n_1} + \dfrac{1}{n_2})}}$

p-value $= \Pr(Z < z)$ or $\Pr(Z > z)$ or $2 \Pr(Z > |z|)$

where $\hat{p}_c$ denotes the *combined* sample proportion (i.e., the proportion of "successes" if the two samples were pooled together as one whole), and Z represents the standard normal distribution

Assumptions:

1. The two samples are independently selected simple random samples from the populations of interest.
2. Both sample sizes are large ($n_1 \geq 30$ and $n_2 \geq 30$ as a rule-of-thumb).

Calculator Features:

1. 2-PropZInt
2. 2-PropZTest

Inference Procedures for a Single Population Mean:

Confidence interval for a population mean μ:

$$\bar{x} \pm t^*_{n-1} \frac{s}{\sqrt{n}},$$

where t^*_{n-1} is the appropriate critical value from the *t*-distribution with $n - 1$ degrees of freedom.

Significance test about a population mean μ:

$$H_0 : \mu = \mu_0$$
$$H_a : \mu < \mu_0, \quad \text{or} \quad H_a : \mu > \mu_0, \quad \text{or} \quad H_a : \mu \neq \mu_0$$

Test statistic: $t = \dfrac{\bar{x} - \mu_0}{s/\sqrt{n}}$

p-value: $\Pr(T_{n-1} < t)$ or $\Pr(T_{n-1} > t)$ or $2 \Pr(T_{n-1} > |t|)$

where μ_0 represents the hypothesized value of the population mean and T_{n-1} represents a t-distribution with $n - 1$ degrees of freedom.

Assumptions:

1. The sample is a simple random sample (SRS) from the population of interest.
2. *Either* that the sample size is large ($n \geq 30$ as a rule-of-thumb) *or* that the population is normally distributed.

Calculator Features:

1. Tinterval
2. TTest

Inference Procedures for Comparing Two Population Means:

Confidence interval for a difference in population means $\mu_1 - \mu_2$:

$$(\bar{x}_1 - \bar{x}_2) \pm t_k^* \sqrt{\frac{s_1^2}{n_1} + \frac{s_2^2}{n_2}} \, ,$$

where t_k^* is the appropriate critical value from the t-distribution with degrees of freedom k, equal to the smaller of $n_1 - 1$ and $n_2 - 1$.

Significance test of equality of two population means μ_1 and μ_2:

$$H_0 : \mu_1 = \mu_2$$
$$H_a : \mu_1 < \mu_2, \quad \text{or} \quad H_a : \mu_1 > \mu_2, \quad \text{or} \quad H_a : \mu_1 \neq \mu_2$$

$$\text{Test statistic: } t = \frac{\bar{x}_1 - \bar{x}_2}{\sqrt{\frac{s_1^2}{n_1} + \frac{s_2^2}{n_2}}}$$

$$\text{p-value} = \Pr(T_k < t) \text{ or } \Pr(T_k > t) \text{ or } 2 \, \Pr(T_k > |t|)$$

where T_k represents a t-distribution with degrees of freedom equal to the smaller of $n_1 - 1$ and $n_2 - 1$.

Assumptions:

1. The two samples are independently selected simple random samples from the populations of interest.
2. *Either* that both sample sizes are large ($n_1 \geq 30$ and $n_2 \geq 30$ as a rule-of-thumb) *or* that both populations are normally distributed.

Calculator Features:

1. 2-SampTInt
2. 2-SampTTest

Table I:

STANDARD NORMAL PROBABILITIES

The table reports the area to the left of the value z under the standard normal curve.

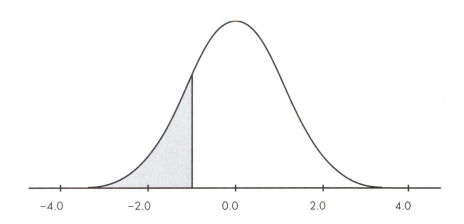

z	.0	.1	.2	.3	.4	.5	.6	.7	.8	.9
−3.4	.0003	.0003	.0003	.0003	.0003	.0003	.0003	.0003	.0003	.0002
−3.3	.0005	.0005	.0005	.0004	.0004	.0004	.0004	.0004	.0004	.0004
−3.2	.0007	.0007	.0006	.0006	.0006	.0006	.0006	.0005	.0005	.0005
−3.1	.0010	.0009	.0009	.0009	.0008	.0008	.0008	.0008	.0007	.0007
−3.0	.0014	.0013	.0013	.0012	.0012	.0011	.0011	.0011	.0010	.0010
−2.9	.0019	.0018	.0018	.0017	.0016	.0016	.0015	.0015	.0014	.0014
−2.8	.0026	.0025	.0024	.0023	.0023	.0022	.0021	.0021	.0020	.0019
−2.7	.0035	.0034	.0033	.0032	.0031	.0030	.0029	.0028	.0027	.0026
−2.6	.0047	.0045	.0044	.0043	.0041	.0040	.0039	.0038	.0037	.0036
−2.5	.0062	.0060	.0059	.0057	.0055	.0054	.0052	.0051	.0049	.0048
−2.4	.0082	.0080	.0078	.0075	.0073	.0071	.0069	.0068	.0066	.0064
−2.3	.0107	.0104	.0102	.0099	.0096	.0094	.0091	.0089	.0087	.0084

z	.0	.1	.2	.3	.4	.5	.6	.7	.8	.9
−2.2	.0139	.0136	.0132	.0129	.0125	.0122	.0119	.0116	.0113	.0110
−2.1	.0179	.0174	.0170	.0166	.0162	.0158	0154	.0150	.0146	.0143
−2.0	.0228	.0222	.0217	0212	.0207	.0202	.0197	.0192	.0188	.0183
−1.9	.0287	.0281	.0274	.0268	.0262	.0256	.0250	.0244	.0239	.0233
−1.8	.0359	.0351	.0344	.0336	.0329	.0322	.0314	.0307	.0301	.0294
−1.7	.0446	.0436	.0427	.0418	.0409	.0401	.0392	.0384	.0375	.0367
−1.6	.0548	.0537	.0526	.0516	.0505	.0495	.0485	.0475	.0465	.0455
−1.5	.0668	.0655	.0643	.0630	.0618	.0606	.0594	.0582	.0571	.0559
−1.4	.0808	.0793	.0778	.0764	.0749	.0735	.0721	.0708	.0694	.0681
−1.3	.0968	.0951	.0934	.0918	.0901	.0885	.0869	.0853	.0838	.0823
−1.2	.1151	.1131	.1112	.1093	.1075	.1057	.1038	.1020	.1003	.0985
−1.1	.1357	.1335	.1314	.1292	.1271	.1251	.1230	.1210	.1190	.1170
−1.0	.1587	.1562	.1539	.1515	.1492	.1469	.1446	.1423	.1401	.1379
−0.9	.1841	.1814	.1788	.1762	.1736	.1711	.1685	.1660	.1635	.1611
−0.8	.2119	.2090	.2061	.2033	.2005	.1977	.1949	.1922	.1894	.1867
−0.7	.2420	.2389	.2358	.2327	.2297	.2266	.2236	.2207	.2177	.2148
−0.6	.2743	.2709	.2676	.2643	.2611	.2578	.2546	.2514	.2483	.2451
−0.5	.3085	.3050	.3015	.2981	.2946	.2912	.2877	.2843	.2810	.2776
−0.4	.3446	.3409	.3372	.3336	.3300	.3264	.3228	.3192	.3156	.3121
−0.3	.3821	.3783	.3745	.3707	.3669	.3632	.3594	.3557	.3520	.3483
−0.2	.4207	.4168	.4129	.4090	.4052	.4013	.3974	.3936	.3897	.3859
−0.1	.4602	.4562	.4522	.4483	.4443	.4404	.4364	.4325	.4286	.4247
−0.0	.5000	.4960	.4920	.4880	.4840	.4801	.4761	.4721	.4681	.4641

The table reports the area to the left of the value z under the standard normal curve.

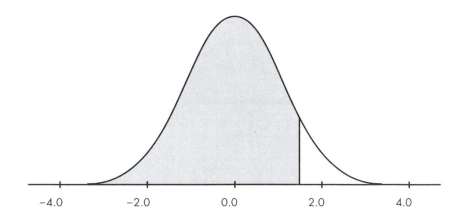

z	.0	.1	.2	.3	.4	.5	.6	.7	.8	.9
0.0	.5000	.5040	.5080	.5120	.5160	.5199	.5239	.5279	.5319	.5359
0.1	.5398	.5438	.5478	.5517	.5557	.5596	.5636	.5675	.5714	.5753
0.2	.5793	.5832	.5871	.5910	.5948	.5987	.6026	.6064	.6103	.6141

z	_0	_1	_2	_3	_4	_5	_6	_7	_8	_9
0.3	.6179	.6217	.6255	.6293	.6331	.6368	.6406	.6443	.6480	.6517
0.4	.6554	.6591	.6628	.6664	.6700	.6736	.6772	.6808	.6844	.6879
0.5	.6915	.6950	.6985	.7019	.7054	.7088	.7123	.7157	.7190	.7224
0.6	.7257	.7291	.7324	.7357	.7389	.7422	.7454	.7486	.7517	.7549
0.7	.7580	.7611	.7642	.7673	.7704	.7734	.7764	.7794	.7823	.7852
0.8	.7881	.7910	.7939	.7967	.7995	.8023	.8051	.8079	.8106	.8133
0.9	.8159	.8186	.8212	.8238	.8264	.8289	.8315	.8340	.8365	.8389
1.0	.8413	.8438	.8461	.8485	.8508	.8531	.8554	.8577	.8599	.8621
1.1	.8643	.8665	.8686	.8708	.8729	.8749	.8770	.8790	.8810	.8830
1.2	.8849	.8869	.8888	.8907	.8925	.8944	.8962	.8980	.8997	.9015
1.3	.9032	.9049	.9066	.9082	.9099	.9115	.9131	.9147	.9162	.9177
1.4	.9192	.9207	.9222	.9236	.9251	.9265	.9279	.9292	.9306	.9319
1.5	.9332	.9345	.9357	.9370	.9382	.9394	.9406	.9418	.9429	.9441
1.6	.9452	.9463	.9474	.9484	.9495	.9505	.9515	.9525	.9535	.9545
1.7	.9554	.9564	.9573	.9582	.9591	.9599	.9608	.9616	.9625	.9633
1.8	.9641	.9649	.9656	.9664	.9671	.9678	.9686	.9693	.9699	.9706
1.9	.9713	.9719	.9726	.9732	.9738	.9744	.9750	.9756	.9761	.9767
2.0	.9773	.9778	.9783	.9788	.9793	.9798	.9803	.9808	.9812	.9817
2.1	.9821	.9826	.9830	.9834	.9838	.9842	.9846	.9850	.9854	.9857
2.2	.9861	.9864	.9868	.9871	.9875	.9878	.9881	.9884	.9887	.9890
2.3	.9893	.9896	.9898	.9901	.9904	.9906	.9909	.9911	.9913	.9916
2.4	.9918	.9920	.9922	.9925	.9927	.9929	.9931	.9932	.9934.	9936
2.5	.9938	.9940	.9941	.9943	.9945	.9946	.9948	.9949	.9951	.9952
2.6	.9953	.9955	.9956	.9957	.9959	.9960	.9961	.9962	.9963	.9964
2.7	.9965	.9966	.9967	.9968	.9969	.9970	.9971	.9972	.9973	.9974
2.8	.9974	.9975	.9976	.9977	.9977	.9978	.9979	.9979	.9980	.9981
2.9	.9981	.9982	.9983	.9983	.9984	.9984	.9985	.9985	.9986	.9986
3.0	.9987	.9987	.9987	.9988	.9988	.9989	.9989	.9989	.9990	.9990
3.1	.9990	.9991	.9991	.9991	.9992	.9992	.9992	.9992	.9993	.9993
3.2	.9993	.9993	.9994	.9994	.9994	.9994	.9994	.9995	.9995	.9995
3.3	.9995	.9995	.9996	.9996	.9996	.9996	.9996	.9996	.9996	.9997
3.4	.9997	.9997	.9997	.9997	.9997	.9997	.9997	.9997	.9997	.9998

Table II:

t-DISTRIBUTION CRITICAL VALUES

The table reports the critical value for which the area to the right is as indicated.

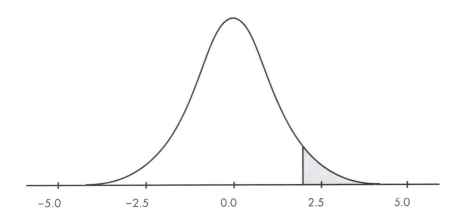

Area to right	0.2	0.1	0.05	0.025	0.01	0.005	0.001	0.0005
Conf. level	60%	80%	90%	95%	98%	99%	99.80%	99.90%
d.f.								
1	1.376	3.078	6.314	12.706	31.821	63.657	318.317	636.607
2	1.061	1.886	2.920	4.303	6.965	9.925	22.327	31.598
3	0.978	1.638	2.353	3.182	4.541	5.841	10.215	12.924
4	0.941	1.533	2.132	2.776	3.747	4.604	7.173	8.610
5	0.920	1.476	2.015	2.571	3.365	4.032	5.893	6.869
6	0.906	1.440	1.943	2.447	3.143	3.708	5.208	5.959
7	0.896	1.415	1.895	2.365	2.998	3.500	4.785	5.408
8	0.889	1.397	1.860	2.306	2.897	3.355	4.501	5.041
9	0.883	1.383	1.833	2.262	2.821	3.250	4.297	4.781
10	0.879	1.372	1.812	2.228	2.764	3.169	4.144	4.587

Area to right	0.2	0.1	0.05	0.025	0.01	0.005	0.001	0.0005
Conf. level	60%	80%	90%	95%	98%	99%	99.80%	99.90%
d.f.								
11	0.876	1.363	1.796	2.201	2.718	3.106	4.025	4.437
12	0.873	1.356	1.782	2.179	2.681	3.055	3.930	4.318
13	0.870	1.350	1.771	2.160	2.650	3.012	3.852	4.221
14	0.868	1.345	1.761	2.145	2.625	2.977	3.787	4.140
15	0.866	1.341	1.753	2.131	2.602	2.947	3.733	4.073
16	0.865	1.337	1.746	2.120	2.583	2.921	3.686	4.015
17	0.863	1.333	1.740	2.110	2.567	2.898	3.646	3.965
18	0.862	1.330	1.734	2.101	2.552	2.878	3.611	3.922
19	0.861	1.328	1.729	2.093	2.539	2.861	3.579	3.883
20	0.860	1.325	1.725	2.086	2.528	2.845	3.552	3.850
21	0.859	1.323	1.721	2.080	2.518	2.831	3.527	3.819
22	0.858	1.321	1.717	2.074	2.508	2.819	3.505	3.792
23	0.858	1.319	1.714	2.069	2.500	2.807	3.485	3.768
24	0.857	1.318	1.711	2.064	2.492	2.797	3.467	3.745
25	0.856	1.316	1.708	2.060	2.485	2.787	3.450	3.725
26	0.856	1.315	1.706	2.056	2.479	2.779	3.435	3.707
27	0.855	1.314	1.703	2.052	2.473	2.771	3.421	3.690
28	0.855	1.313	1.701	2.048	2.467	2.763	3.408	3.674
29	0.854	1.311	1.699	2.045	2.462	2.756	3.396	3.659
30	0.854	1.310	1.697	2.042	2.457	2.750	3.385	3.646
40	0.851	1.303	1.684	2.021	2.423	2.704	3.307	3.551
50	0.849	1.299	1.676	2.009	2.403	2.678	3.261	3.496
60	0.848	1.296	1.671	2.000	2.390	2.660	3.232	3.460
80	0.846	1.292	1.664	1.990	2.374	2.639	3.195	3.416
100	0.845	1.290	1.660	1.984	2.364	2.626	3.174	3.391
500	0.842	1.283	1.648	1.965	2.334	2.586	3.107	3.310
Infinity	0.842	1.282	1.645	1.960	2.326	2.576	3.090	3.291

Appendix A:

LISTINGS OF HYPOTHETICAL DATA SETS

Many of the hypothetical data sets used in the text present graphical displays without listing the actual data values. These values are necessary in some calculations, so they are presented below.

Activities 3-1: Hypothetical Exam Scores; 4-3: Properties of Averages; 5-2: Properties of Measures of Spread

Class A	Class B	Class C	Class D	Class E	Class F	Class G	Class H	Class I
89	71	66	69	71	56	37	33	37
73	71	54	72	71	61	58	40	57
73	52	55	70	52	67	56	38	55
75	74	61	71	74	57	47	52	16
88	74	59	72	74	81	45	76	51
95	72	62	74	72	68	40	62	60
81	69	59	72	69	75	42	38	27
73	75	56	68	75	72	63	73	65
76	63	57	79	63	91	51	88	66
73	69	58	70	69	71	60	54	53
75	65	50	70	65	66	66	59	52
80	65	66	71	65	96	32	39	67
81	63	59	72	63	83	79	33	42
92	72	50	70	72	62	47	57	59
98	75	61	72	75	78	73	51	47
85	60	66	71	60	71	29	46	65
70	65	47	68	65	61	64	41	51
77	73	46	73	73	68	49	54	47
85	68	60	72	68	72	51	51	33
83	66	62	70	66	72	63	43	56
88	68	63	72	68	60	62	62	25

Class A	Class B	Class C	Class D	Class E	Class F	Class G	Class H	Class I
74	78	60	70	78	64	44	45	63
84	67	77	66	67	71	30	51	66
90	77	67	72	77	58	43	51	51
81	63	55	73	63	72	57	49	38
81	70	51	69	70	50	70	68	57
83	63	65	66	63	70	33	39	22
68	66	60	72	66	82	58	57	31
91	59	55	67	59	73	61	59	46
82	61	72	68	61	72	51	49	52
70	74	63	71	74	65	65	84	37
78	75	68	72	75	90	29	37	45
77	72	65	72	72	55	76	52	40
89	53	62	67	53	70	50	34	59
73	69	66	68	69	80	71	33	50
81	66	68	70	66	80	47	75	27
79	76	61	75	76	72	52	62	57
83	57	55	71	57	83	37	37	55
73	76	47	71	76	73	25	45	53
87	67	72	76	67	55	38	51	55
72	73	52	71	73	81	56	56	46
81	73	74	73	73	73	39	50	50
97	64	62	65	64	71	68	45	24
77	89	70	74	89	56	53	66	57
79	78	60	72	78	71	54	43	47
76	68	41	68	68	49	47	39	53
71	73	68	72	73	50	32	46	59
79	71	73	68	71	77	50	44	61
85	74	52	69	74	65	31	56	59
85	82	47	71	82	77	60	43	42
77	70	46	70	70	67	28	47	68
80	70	53	73	70	71	37	57	58
83	57	63	73	57	55	77	53	56
85	72	66	73	72	79	63	39	32
76	63	72	75	63	67	49	39	61
83	73	56	62	73	79	49	48	35
73	77	45	70	77	49	58	48	57
72	76	59	69	76	56	58	42	52
79	64	48	73	64	67	76	36	49
90	69	52	64	69	74	60	39	62
78	65	67	74	65	46	22	56	58
90	62	78	70	62	67	31	52	42
79	83	64	74	83	69	29	70	54
76	67	60	70	67	76	47	50	65
64	66	51	67	66	76	46	63	65
68	56	56	68	56	75	45	42	46
80	78	66	68	78	85	34	52	68
88	73	53	64	73	45	46	50	66
87	76	65	71	76	86	21	57	57

Class A	Class B	Class C	Class D	Class E	Class F	Class G	Class H	Class I
85	70	74	70	70	81	42	36	25
82	72	62	70	72	74	54	38	63
85	59	61	68	59	71	78	44	50
79	66	62	66	66	88	47	40	58
90	83	66	70	83	76	58	42	61
88	73	60	69	73	79	49	38	44
87	79	62	70	79	95	53	57	40
81	65	67	68	65	72	45	51	56
87	61	59	77	61	80	52	68	38
82	64	61	70	64	61	42	36	52
87	78	68	68	78	72	28	74	66
81	57	62	65	57	85	38	68	61
81	69	76	71	69	77	38	51	56
79	74	53	72	74	77	52	81	64
82	64	61	73	64	76	32	48	51
91	64	57	66	64	40	35	73	59
75	83	61	70	83	96	47	44	60
64	73	61	74	73	59	43	37	67
82	70	66	69	70	69	44	52	65
89	79	54	72	79	68	46	72	36
86	71	65	71	71	53	37	40	54
73	73	57	72	73	72	53	46	43
80	70	57	73	70	70	73	41	48
68	72	48	72	72	75	67	36	52
73	77	59	70	77	84	66	36	60
72	68	67	74	68	76	57	91	52
84	66	58	71	66	81	52	61	51
74	71	56	68	71	71	69	49	40
85	55	61	70	55	55	64	41	55
76	72	60	70	72	99	53	41	57
83	59	60	74	59	86	46	33	52

Activities 3-6 and 5-5: Hypothetical Manufacturing Processes

Process A	Process B	Process C	Process D
11.5378	12.3179	11.9372	12.1122
11.4454	11.6369	11.4516	11.8743
11.4973	11.9653	11.6774	11.9922
11.4806	11.9294	11.6848	11.9810
11.5251	12.0735	11.7001	12.0227
11.4221	11.7086	11.5958	11.9032
11.5107	12.0805	11.7025	12.0116
11.5657	12.5956	12.1176	12.1953
11.4678	12.0636	11.8339	12.0202
11.4818	11.9256	11.6309	11.9640
11.4988	12.2369	11.9200	12.0744

Process A	Process B	Process C	Process D
11.4165	11.4654	11.3773	11.8257
11.5538	12.3717	11.9023	12.1138
11.5097	12.3041	11.9396	12.0898
11.4662	11.9887	11.7716	11.9980
11.4398	11.6219	11.4825	11.8800
11.4478	11.7303	11.6008	11.9260
11.5635	12.3644	11.8640	12.1089
11.5397	12.1163	11.7297	12.0446
11.4753	11.8964	11.5965	11.9472
11.5313	12.1731	11.7979	12.0604
11.5102	11.9675	11.6818	12.0044
11.5348	12.1768	11.7582	12.0501
11.4574	11.5570	11.3846	11.8618
11.5077	11.9915	11.6969	12.0073
11.4882	12.0150	11.7551	12.0107
11.5391	12.4587	12.0131	12.1386
11.4908	12.0230	11.7715	12.0182
11.5439	12.3769	11.9613	12.1253
11.4235	11.6453	11.5309	11.8827
11.4576	11.8803	11.6867	11.9627
11.4208	11.6528	11.5561	11.8888
11.4408	11.7646	11.6197	11.9266
11.5007	12.1889	11.8806	12.0628
11.4828	11.7945	11.5535	11.9390
11.4839	12.0097	11.7472	12.0045
11.5457	12.1448	11.7488	12.0559
11.5636	12.4386	11.9927	12.1520
11.4733	11.8538	11.6834	11.9745
11.4608	11.5233	11.3408	11.8500
11.4485	11.7976	11.6320	11.9369
11.4585	11.6921	11.4837	11.8958
11.4806	12.1321	11.8958	12.0513
11.5222	12.0173	11.6354	11.9988
11.4636	11.7704	11.5487	11.9216
11.5088	12.1286	11.7875	12.0384
11.4551	11.9278	11.7457	11.9803
11.5165	12.3054	11.9466	12.0977
11.5358	12.1318	11.7388	12.0444
11.5402	12.4835	12.0710	12.1588

Activity 5-14: Guessing Standard Deviations

Data A	Data B	Data C	Data D
54.7323	200.2330	1.0614	−3.7427
58.0655	182.6690	1.0890	−8.3320
75.4949	152.5970	0.9694	−7.8994

Data A	Data B	Data C	Data D
70.8519	176.1100	0.9964	−6.6365
67.5594	284.0310	0.9779	−8.5103
82.0604	210.7530	0.9920	−11.3988
67.3449	135.7440	1.0129	−2.9734
73.9735	167.4280	0.9910	−3.6779
59.9569	206.9280	0.9443	−0.1072
64.0481	127.0890	0.9710	−3.1872
70.4145	142.6490	1.0321	−3.3980
70.5806	225.1930	0.9837	−2.4381
70.5234	134.7350	0.9651	5.4287
59.6053	191.7320	1.0173	−1.4514
63.0206	177.6910	1.0044	−5.5888
57.1386	142.7390	0.9559	−3.4043
66.7414	172.8910	0.9539	−4.8423
66.2912	238.4620	1.0476	−4.4200
39.4021	343.0530	0.9888	−0.5386
70.4660	314.2950	1.0985	−9.1686
54.9322	182.8570	0.9819	−8.6683
70.0909	244.5870	0.9856	−14.3791
49.2586	236.8060	0.9920	−8.4293
62.5470	268.3330	0.9751	−14.4531
54.5510	244.9190	1.0732	−4.0124
78.2332	169.8710	1.0083	−2.9723
75.0430	226.4780	0.9905	−4.6135
74.6654	269.1320	0.8907	−5.1762
76.7585	203.2910	0.9641	−12.1286
83.1006	201.5240	1.0290	−12.3931
66.9678	167.4570	0.9312	0.1197
51.2948	102.8890	0.9725	−4.6073
79.2687	205.5250	1.1043	1.8484
68.5335	200.3780	1.0372	−1.4457
51.1347	220.1900	1.0607	−15.2446
53.6251	265.0840	1.0061	−1.9615
73.7969	221.6660	0.9644	−8.9919
61.4841	214.0470	0.9515	−6.0234
76.7655	263.5450	0.9307	−5.9924
54.0086	234.6190	1.0096	−0.1938
66.8219	205.1490	0.9744	−5.5941
55.5908	148.4350	0.9623	−8.6759
70.0916	260.8790	0.9661	−8.7433
63.3856	263.4330	0.9308	−2.1252
64.9589	182.7400	0.9820	−1.3464
74.9712	115.6500	1.0369	−11.2186
63.5829	270.8650	0.9952	−5.0162
65.8217	245.4750	1.0101	−0.9512
76.4025	126.9230	0.9952	−5.3799
53.4284	198.1700	0.9717	1.6207
64.9588	231.7590	0.9647	−2.7879

Data A	Data B	Data C	Data D
46.9573	253.3560	0.9833	−4.8769
62.2297	241.4200	0.9930	−5.1715
71.0247	242.9490	1.0570	−6.1173
69.9688	261.1060	0.9557	0.5181
69.0292	172.6260	0.8707	−2.2005
53.4182	207.3680	1.0125	−1.2335
69.6185	179.3460	1.0678	−11.8271
57.9149	234.7990	1.0749	−3.4148
68.8708	174.7420	0.9809	−9.6962
68.5252	129.6640	0.9940	−9.9124
76.7544	95.4220	1.0030	−11.3669
64.6388	133.3370	0.9740	−4.3015
68.7167	263.5060	1.0385	−10.4150
54.8736	162.8340	0.9938	−9.1377
48.9441	251.2650	0.8861	−13.9931
58.3347	193.2390	1.0176	−16.0001
63.1308	223.1980	1.0385	−9.6562
38.8426	268.0770	1.0698	−8.9456
49.4594	130.5850	1.0769	3.9791
66.3701	228.6850	0.9678	−2.3438
69.3799	183.8290	1.0098	−9.5794
60.4799	142.1750	0.9816	−0.6389
73.7690	201.2340	0.9804	−2.6748
61.7042	266.1770	1.0372	2.1504
78.7420	240.5090	0.9559	−7.3905
41.3535	320.8660	1.0449	−5.1572
64.8503	177.7580	1.0209	−3.6856
60.8477	204.0690	0.9932	−13.5580
70.9215	119.6970	0.9964	−8.8984
58.0815	160.1130	1.0225	3.3616
63.3596	130.8600	0.9923	−8.0555
73.5292	204.0180	0.8370	2.6732
58.9281	226.8270	1.0737	−1.5957
65.5045	65.3540	0.9700	−0.4293
57.2830	265.8720	1.1029	−5.3904
54.3327	182.7050	1.0509	−0.8345
61.8720	197.8660	0.9651	−6.4203
76.2123	231.3630	1.0127	−15.0415
58.2271	146.0190	1.1274	−2.1868
88.1693	149.2930	1.0328	−4.1268
65.4506	206.2060	0.9670	−0.1929
60.7583	208.3380	0.9833	−6.0495
56.7109	272.6800	0.9848	−1.9930
80.8921	193.3570	1.0250	−10.1710
54.2190	238.9720	1.0252	−4.2453
66.8669	185.3370	1.0170	−0.7348
71.8376	168.8880	0.9294	−14.4362
59.6969	222.3630	1.0116	−6.9096
59.4178	172.1620	1.0410	−5.9749

Activities 7-2: Guess the Association; 8-1: Properties of Correlation

Exam1 A	Exam2 A	Exam1 B	Exam2 B	Exam1 C	Exam2 C
72	78	78	78	69	72
79	74	75	63	84	55
69	67	88	77	55	83
76	77	73	78	78	62
56	60	85	68	67	76
60	60	78	73	81	61
64	65	76	67	57	84
78	70	68	70	57	80
78	71	76	60	72	67
68	74	67	75	83	56
74	77	76	67	61	79
66	79	60	57	77	62
74	82	58	52	73	68
83	85	66	66	68	74
82	77	66	78	64	77

Exam1 D	Exam2 D	Exam1 E	Exam2 E	Exam1 F	Exam2 F
79	62	99	96	55	68
67	61	59	56	57	61
62	70	75	76	64	68
64	78	58	58	64	70
82	59	82	83	64	59
81	69	77	77	73	64
71	68	95	95	64	77
63	67	58	61	67	56
67	65	65	65	85	56
63	75	68	69	68	57
57	75	68	66	60	64
59	74	59	58	69	66
65	74	89	85	73	59
80	58	56	59	59	70
68	67	84	80	67	60

Exam1 G	Exam2 G	Exam1 H	Exam2 H	Exam1 I	Exam2 I	Exam1 J	Exam2 J
49	95	49	54	12	17	37	39
52	81	52	57	52	62	44	41
55	69	55	60	55	52	32	33
58	59	58	62	58	73	37	34
61	51	61	65	61	69	35	41
64	45	64	68	64	71	41	45
67	41	67	70	67	72	42	33
70	39	70	73	70	80	45	36

Exam1 G	Exam2 G	Exam1 H	Exam2 H	Exam1 I	Exam2 I	Exam1 J	Exam2 J
73	41	73	76	73	58	88	78
76	45	76	78	76	60	72	74
79	51	79	81	79	67	75	81
82	59	82	84	82	52	81	77
85	69	85	87	85	76	82	94
88	81	88	89	88	69	71	84
91	95	99	12	91	70	82	87

Activity 15-15: Random Normal Data

Exam 1	Exam 2	Exam 3	Exam 4	Exam 5
62	58	66	74	90
67	57	89	88	82
66	66	78	87	78
65	90	75	89	79
59	59	78	86	86
75	62	90	74	85
71	57	65	79	86
78	56	74	88	76
74	74	69	88	81
64	89	83	79	88
65	67	66	80	87
75	69	89	83	72
59	57	78	82	72
71	93	89	88	79
78	70	69	88	78
61	58	67	57	81
67	89	83	64	76
68	65	65	88	81
85	57	90	89	78
71	63	69	89	85
93	62	68	85	76
83	59	90	82	73
61	59	71	86	81
73	73	75	83	78
69	59	65	85	83
69	59	65	89	75
77	57	73	85	89
70	58	79	68	84
86	56	83	71	69
69	63	69	80	79

Activities 24-2, 24-5, and 24-10: Hypothetical Sleeping Times

Sample 1	Sample 2	Sample 3				Sample 4	
8.5	6.1	7.1	9.5	7.1	8.7	9.3	8.7
8.4	10.1	8.3	7.1	7.4	9.4	5.3	7.4
7.8	5.7	7.1	7.5	7.4	6.6	7.3	6.3
8.3	6.2	7.9	7.9	7.8	6.0	6.7	5.9
7.1	8.2	7.5	6.4	6.2	6.9	5.8	10.0
6.4	6.3	6.2	6.2	8.6	9.9	4.7	6.5
8.0	7.6	8.2	7.5	8.4	6.3	5.6	8.6
6.9	7.0	8.7	7.7	6.6	8.9	5.9	7.7
6.4	9.4	8.5	7.6	8.1	10.1	9.4	9.0
8.2	9.4	7.6	8.8	7.1	9.6	7.6	7.9

Activities 26-1 and 26-4: Hypothetical Commuting Times

Alex 1	Alex 2	Barb 1	Barb 2	Carl 1	Carl 2
19.3	23.7	16.4	24.4	23.7	27.7
20.5	24.5	17.6	27.1	24.3	28.5
23.0	27.7	20.2	32.0	25.5	29.1
25.8	30.0	21.0	34.0	26.8	30.2
28.0	31.9	23.8	35.1	28.0	32.1
28.8	32.5	26.4	36.1	28.4	32.4
30.6	32.6	30.4	37.8	29.3	34.0
32.1	35.5	30.6	38.3	30.1	34.6
33.5	38.7	31.6	41.3	30.6	35.3
38.4	42.9	32.0	43.9	33.3	36.1

Donna 1				Donna 2			
10.1	24.9	28.4	32.4	20.4	27.7	32.4	36.4
13.5	25.0	28.5	33.3	22.4	28.6	32.5	36.6
20.9	25.8	28.9	33.4	24.2	28.9	33.8	37.2
21.7	26.0	29.0	33.8	24.4	28.9	33.9	37.3
23.7	26.1	29.3	33.8	24.4	29.4	34.0	37.5
22.8	26.5	29.8	34.0	25.0	29.5	34.9	38.0
23.0	26.6	30.1	35.3	25.4	30.4	35.4	39.7
23.5	26.7	30.0	36.6	25.6	30.4	35.5	41.5
24.0	27.1	30.2	39.5	26.2	31.0	35.5	44.4
24.8	28.0	33.2	39.9	27.2	31.2	36.1	46.2

Appendix B:

SOURCES FOR DATA SETS

Topic 1: Data and Variables I

- *Activity 1-5:* data on physicians' gender are from *The 1995 World Almanac and Book of Facts*, p. 966.
- *Activity 1-9:* data on sports' hazardousness are reported in the November 12, 1993, issue of the *Harrisburg Evening-News*.

Topic 2: Data and Variables II

- *Activity 2-4:* data on tennis simulations are a random sample from the analysis in "Computer Simulation of a Handicap Scoring System for Tennis," by Allan Rossman and Matthew Parks, *Stats: The Magazine for Students of Statistics*, 10, 1993, pp. 14–18.
- *Activity 2-7:* data on Broadway shows are from a June 1993 issue of *Variety* magazine.

Topic 3: Displaying and Describing Distributions

- *Activity 3-2:* data on British rulers are from *The 1995 World Almanac and Book of Facts*, pp. 534–535.
- *Activity 3-3:* data on college tuitions are from the October 23, 1991, issue of *The Chronicle of Higher Education*.
- *Activity 3-7:* data on marriage ages were gathered by Matthew Parks at the Cumberland County (PA) Courthouse in June–July 1993.

- *Activity 3-8:* data on Hitchcock films were gathered at a Blockbuster Video store in Carlisle, PA, in January 1995.
- *Activity 3-9:* data on dinosaur heights are from *Jurassic Park* by Michael Crichton, Ballantine Books, 1990, p. 165.
- *Activity 3-10:* data on movies' incomes are from issues of *Variety* magazine in the summer of 1993.
- *Activity 3-14:* data on Perot votes are from *The 1993 World Almanac and Book of Facts*, p. 73.

Topic 4: Measures of Center

- *Activity 4-1:* data on Supreme Court service are from *The 1995 World Almanac and Book of Facts*, p. 86.
- *Activity 4-3:* data on cancer pamphlets are from "Readability of Educational Materials for Cancer Patients," by Thomas H. Short, Helene Moriarty, and Mary Cooley, *Journal of Statistics Education*, 3, 1995.
- *Activity 4-6:* data on planetary measurements are from *The 1993 World Almanac and Book of Facts*, p. 251.
- *Activity 4-7:* data on the Supreme Court are from *The 1995 World Almanac and Book of Facts*, p. 85.

Topic 5: Measures of Spread

- *Activity 5-6:* data on climate are from Tables 368–375 of the *1992 Statistical Abstract of the United States*.

Topic 6: Comparing Distributions

- *Activity 6-1:* data on shifting populations are from *The 1995 World Almanac and Book of Facts*, p. 379.
- *Activity 6-2:* data on golfers' winnings are from the February 1991 issue of *Golf* magazine.
- *Activity 6-7:* data on automobile thefts are from Table 209 of the *1992 Statistical Abstract of the United States*.
- *Activity 6-8:* data on lifetimes are from *The 1991 World Almanac and Book of Facts*, pp. 336–374.
- *Activity 6-11:* data on governors' salaries are from *The 1995 World Almanac and Book of Facts*, pp. 94–97.

- *Activity 6-13:* data on cars are from *Consumer Reports' 1995 New Car Yearbook.*
- *Activity 6-15:* data on mutual funds are from the January 7, 1994, issue of *The Wall Street Journal.*
- *Activity 6-16:* data on *Star Trek* episodes are from *Entertainment Weekly's Special Star Trek Issue,* 1994.

Topic 7: Graphical Displays of Association

- *Activity 7-4:* data on fast food sandwiches are from a 1993 Arby's nutritional brochure.
- *Activity 7-5:* data on O-ring failures are from "Lessons Learned from Challenger: A Statistical Perspective," by Siddhartha R. Dalal, Edward B. Folkes, and Bruce Hoadley, *Stats: The Magazine for Students of Statistics,* 2, 1989, pp. 14–18.
- *Activity 7-7:* data on airfares are from the January 8, 1995, issue of the *Harrisburg Sunday Patriot-News* and from the Delta Air Lines worldwide timetable guide effective December 15, 1994.
- *Activity 7-11:* data on alumni donations are from the 1991–1992 Annual Report of Dickinson College.
- *Activity 7-12:* data on peanut butter are from the September 1990 issue of *Consumer Reports* magazine.
- *Activity 7-13:* data on SAT scores are from the August 31, 1992, issue of the *Harrisburg Evening-News.*

Topic 8: Correlation Coefficient

- *Activity 8-2:* data on televisions and life expectancy are from *The 1993 World Almanac and Book of Facts,* pp. 727–817.
- *Activity 8-14:* data on "Top Ten" rankings are from *An Altogether New Book of Top Ten Lists* by David Letterman et al., Pocket Books, 1991, p. 44.

Topic 9: Least Squares Regression I

- *Activity 9-7:* data on basketball salaries are from the June 24, 1992, issue of *USA Today.*

Topic 10: Least Squares Regression II

- *Activity 10-1:* data on gestation and longevity are from *The 1993 World Almanac and Book of Facts*, p. 676.
- *Activity 10-6:* data on college enrollments are a sample from *The 1991 World Almanac and Book of Facts*, pp. 214–239.

Topic 11: Relationships with Categorical Variables

- *Activity 11-2:* data on age and ideology are from *Vital Statistics on American Politics* by Harold W. Stanley and Richard G. Niemi, CQ Press, 1988, pp. 130–131.
- *Activity 11-3:* data on AZT and HIV are reported in the March 7, 1994, issue of *Newsweek*.
- *Activity 11-6:* data on toy advertising were supplied by Dr. Pamela Rosenberg.
- *Activity 11-11:* data on living arrangements are from *The 1995 World Almanac and Book of Facts*, p. 961.
- *Activity 11-12:* data on Civil War generals are summarized from *Civil War Generals* by James Spencer, Greenwood Press, 1986, pp. 121–138.
- *Activity 11-13:* data on graduate admissions are from "Is There Sex Bias in Graduate Admissions?" by P. J. Bickel and J. W. O'Connell, *Science*, 187, pp. 398–404.
- *Activity 11-14:* data on baldness and heart disease are from "A Case-Control Study of Baldness in Relation to Myocardial Infarction in Men," by Samuel M. Lasko et al., *Journal of the American Medical Association*, 269, 1993, pp. 998–1003.

Topic 12: Random Sampling

- *Activity 12-1:* data regarding the Elvis poll are reported in the August 18, 1989, issue of the *Harrisburg Patriot-News*. The data on the *Literary Digest* poll are from "Why the Literary Digest Poll Failed," by Peverill Squire, *Public Opinion Quarterly*, 52, 1988, pp. 125–133.
- *Activity 12-2:* data on U.S. Senators are from *The 1995 World Almanac and Book of Facts*, pp. 76–77.
- *Activity 12-5:* data on polls regarding emotional support are reported in *The Superpollsters* by David W. Moore, Four Walls Eight Windows Publishers, p. 19.

- *Activity 12-6:* data on alternative medicine are from a March 1994 issue of *Self* magazine.
- *Activity 12-7:* data on courtroom cameras are reported in the October 4, 1994, issue of the *Harrisburg Evening-News.*

Topic 13: Sampling Distributions I: Confidence

- *Activity 13-4:* data on Presidential voting are from *The 1995 World Almanac and Book of Facts,* p. 601.
- *Activity 13-7:* data on American moral decline are from the June 13, 1994, issue of *Newsweek.*
- *Activity 13-8:* data on cat households are from Table 392 of the *1992 Statistical Abstract of the United States.*

Topic 15: Normal Distributions

- *Activity 15-9:* data on family lifetimes were supplied by Anthony Kapolka.

Topic 16: Central Limit Theorem

- *Activity 16-11:* data on non-English speakers are from *The 1995 World Almanac and Book of Facts,* p. 600.

Topic 17: Confidence Intervals I

- *Activity 17-10:* data on television characters were reported in the June 28, 1994, issue of *USA Today.*
- *Activity 17-12:* data on charitable contributions are from Table 603 of the *1992 Statistical Abstract of the United States.*

Topic 18: Confidence Intervals II

- *Activity 18-5:* data on dissatisfaction with Congress are from the October 27, 1994, issue of *USA Today.*
- *Activity 18-12:* data on marital problems are from "Why Does Military Combat Experience Adversely Affect Marital Relations?" by Cynthia Gimbel and Alan Booth, *Journal of Marriage and the Family,* 56, 1994, pp. 691–703.
- *Activity 18-13:* data on jury representativeness are from "Statistical Evidence of Discrimination," by David Kaye, *Journal of the American Statistical Association,* 77, 1982, pp. 773–783.

Topic 19: Tests of Significance I

- *Activity 19-5:* data on teacher hiring are reported in *Statistics for Lawyers,* by Michael O. Finkelstein and Bruce Levin, Springer-Verlag, 1990, pp. 161–162.

Topic 21: Designing Experiments

- *Activity 21-7:* data on UFO sightings are from "Close Encounters: An Examination of UFO Experiences," by Nicholas P. Spanos, Patricia A. Cross, Kirby Dickson, and Susan C. DuBreuil, *Journal of Abnormal Psychology,* 102, 1993, pp. 624–632.
- *Activity 21-8:* data on Mozart music are reported in the October 14, 1993, issue of *Nature.*
- *Activity 21-11:* data on parolees are reported in the October 6, 1993, issue of *The New York Times,* p. B10.

Topic 22: Comparing Two Proportions I

- *Activity 22-5:* data on wording of surveys are from Table 8.1 of *Questions and Answers in Attitude Surveys* by Howard Schuman and Stanley Presser, Academic Press, 1981.
- *Activity 22-6:* data on wording of surveys are from Tables 11.2, 11.3, and 11.4 of *Questions and Answers in Attitude Surveys* by Howard Schuman and Stanley Presser, Academic Press, 1981.
- *Activity 22-7:* data on smoking policies were supplied by Janet Meyer.

Topic 23: Comparing Two Proportions II

- *Activity 23-2:* data on alcohol habits are from "Boozing and Brawling on Campus: A National Study of Violent Problems Associated with Drinking over the Past Decade," by Ruth C. Enge and David J. Hanson, *Journal of Criminal Justice*, 22, 1994, pp. 171–180.
- *Activity 23-6:* data on the BAP Study are from "The Epidemiology of Bacillary Angiomatosis and Bacillary Peliosis," by Jordan W. Tappero et al., *Journal of the American Medical Association*, 269, 1993, pp. 770–775.
- *Activity 23-8:* data on television sex are reported in *Hollywood vs. America* by Michael Medved, HarperCollins Publishers, 1992, pp. 111–112.
- *Activity 23-9:* data on heart surgery are from a study by the Pennsylvania Health Care Cost Containment Council reported in the November 20, 1992, issue of the *Harrisburg Patriot-News.*
- *Activity 23-10:* data on employment discrimination are reported in *Statistics for Lawyers* by Michael O. Finkelstein and Bruce Levin, Springer-Verlag, 1990, p. 123.
- *Activity 23-12:* data on kids' smoking are reported in the October 4, 1994, issue of the *Harrisburg Patriot-News.*

Topic 26: Comparing Two Means

- *Activity 26-5:* data on baseball games were collected by Matthew Parks in June 1992.
- *Activity 26-9:* data on classroom attention are from "Learning What's Taught: Sex Differences in Instruction," by Gaea Leinhardt, Andrea Mar Seewald, and Mary Engel, *Journal of Educational Psychology*, 71, 1979, pp. 432–439.

Index

DATE DUE

GAYLORD			PRINTED IN U.S.A.